How To Weld

Todd Bridigum

The Complete Street Machine Library™

This edition published by National Street Machine Club by arrangement with Motorbooks, a division of Quayside Publishing.

ISBN-13: 978-1-58159-439-3

Cover photo credits:
Front cover, top image: *Monte Swan*
Front cover, main image: *Mark Simpson*
Back cover, all images: *Monte Swan*

About the author
Todd Bridigum is a long-time, American Welding Society–certified (Certified Welding Inspector, Certified Welding Educator, and Certified Welder to AWS D1.1 Structural Welding Code – Steel) welding instructor at Minneapolis Community Technical College. He lives in Minneapolis, Minnesota.

Editor: James Manning Michels
Designer: Danielle Smith

Printed in Singapore

National Street Machine Club
12301 Whitewater Drive
Minnetonka, MN 55343
www.streetmachineclub.com

CONTENTS

INTRODUCTION

Many books have been written about welding, so when I took up the challenge to write one, I wanted to set it apart from the others. Some of the published books have a lot of technical content, but are difficult for the non-professional to read. Others specialize in one or two welding processes, are outdated, or fall short in explaining how to weld. As a welding instructor I've taught hundreds of students how to weld with varying degrees of success. In my experience some people have a natural talent for it; for others it is more difficult to learn. Only a few just didn't get it or were too afraid of fire, but I could count them on one hand.

The next question is, can you learn how to weld from a book or do you need an instructor's help? The role I play in the classroom is that of diagnostician. In other words, after someone knows how to operate the equipment safely and has done some basic exercises then it is a matter of figuring out what they are doing wrong and making a correction.

Troubleshooting is the key to good welding. Sometimes it is something wrong with the equipment and/or materials being used, which I always look at first and is easy to fix. If it is not the materials or equipment then it is operator error—something is wrong with the student's welding technique. This is a far more difficult problem to correct because it has to do with an individual's hand/eye coordination, body position, perception of distances and angles, and old habits. Without an instructor telling you what to change, it is more difficult to learn how to weld, but not impossible.

In this book I strive to provide useful information for you to work safely and troubleshoot your own problems, yet I would strongly advise taking a class to get more welding time and access to a variety of welding and fabrication equipment. Bear in mind that not all instructors are created equal. If you can get recommendations from former welding students or people who work at your local welding supplier, you may find an instructor who can help you a great deal.

ARC WELDING (AW)

arc stud welding	SW
atomic hydrogen welding	AHW
bare metal arc welding	BMAW
carbon arc welding	CAW
gas carbon arc welding	CAW-G
shielded carbon arc welding	CAW-S
twin carbon arc welding	CAW-T
electrogas welding	EGW
flux cored arc welding	FCAW
gas-shielded flux cored arc welding	FCAW-G
self-shielded flux cored arc welding	FCAW-S
gas metal arc welding	GMAW
pulsed gas metal arc welding	GMAW-P
short circuit gas metal arc welding	GMAW-S
gas tungsten arc welding	GTAW
pulsed gas tungsten arc welding	GTAW-P
magnetically impelled arc welding	MIAW
plasma arc welding	PAW
shielded metal arc welding	SMAW
submerged arc welding	SAW
series submerged arc welding	SAW-S

RESISTANCE WELDING (RW)

flash welding	FW
pressure-controlled resistance welding	RW-PC
projection welding	PW
resistance seam welding	RSEW
high-frequency seam welding	RSEW-HF
induction seam welding	RSEW-I
mash seam welding	RSEW-MS
resistance spot welding	RSW
upset welding	UW
high-frequency	UW-HF
induction	UW-I

SOLDERING (S)

dip soldering	DS
furnace soldering	FS
induction soldering	IS
infrared soldering	IRS
iron soldering	INS
resistance soldering	RS
torch soldering	TS
ultrasonic soldering	USS
pressure gas soldering	WS

WELDING AND JOINING PROCESSES

SOLID STATE WELDING (SSW)

coextrusion welding	CEW
cold welding	CW
diffusion welding	DFW
hot isostatic pressure welding	HIPW
explosion welding	EXW
forge welding	FOW
friction welding	FRW
direct drive friction welding	FRW-DD
friction stir welding	FSW
inertia friction welding	FRW-I
hot pressure welding	HPW
roll welding	ROW
ultrasonic welding	USW

OXYFUEL GAS WELDING (OFW)

air acetylene welding	AAW
oxyacetylene welding	OAW
oxyhydrogen welding	OHW
pressure gas welding	PGW

BRAZING (B)

block brazing	BB
diffusion brazing	DFB
dip brazing	DB
exothermic brazing	EXB
furnace brazing	FB
induction brazing	IB
infrared brazing	IRB
resistance brazing	RB
torch brazing	TB
twin carbon arc brazing	TCAB

OTHER WELDING AND JOINING

adhesive bonding	AB
braze welding	BW
arc braze welding	ABW
carbon arc braze welding	CABW
electron beam braze welding	EBBW
exothermic braze welding	EXBW
flow brazing	FLB
flow welding	FLOW
laser beam braze welding	LBBW
electron beam welding	EBW
high vacuum	EBW-HV
medium vacuum	EBW-MV
nonvacuum	EBW-NW
electroslag welding	ESW
consumable guide electroslag welding	ESW-CG
induction welding	IW
laser beam welding	LBW
percussion welding	PEW
thermite welding	TW

AWS A3.0:2001, Figure 54a, Reproduced with permission from the American Welding Society (AWS), Miami, FL USA

CHAPTER 1
HISTORY AND PROCESS OVERVIEW

Here is a fun way to think about how important welding and metal fabrication are to our standard of living today. What if all the welds in the world came apart at one time and welders weren't available to repair them? Society as we know it would fall apart, or at least electricity would go away. Power plants are held together by welds and maintained by pipefitters, ironworkers, boilermakers, and machinists. All of us rely on electricity and on those skilled professionals for our standard of living. What about wind and solar panels? Each of these relies on manufacturing processes and welding to be fabricated and installed.

I believe metal fabrication is the single largest influence on how the world looks today, from the cities to the country and even into space. Today we tend to think in the short term, which is understandable since our computers, cell phones, cars, and televisions seem to be outdated or obsolete every few years. If we back up and look at the larger picture in the context of welding, the technology available to work and join metals is extraordinary.

Humankind moved from the Stone Age into the Bronze Age about 5,500 years ago. Copper was the first metal to be worked because it exists in its pure form in nature, and using it didn't require separating iron from ore. About 3,200 years ago, people discovered how to separate iron from rock and the Iron Age began. Forge welding, the first metal-joining process, was developed during the Iron Age. The people able to transform rock into usable iron tools through forge welding, known to us today as blacksmiths, were thought to possess magical powers a few millennia ago. It took a lot of time, work, and skill to fabricate metal, and for thousands of years that is *all* we had.

Then, in the late 1800s and early 1900s, gas, arc, and resistance welding were discovered. These breakthroughs, combined with Bessemer's furnace (which enabled industry to make large quantities of high-quality steel) led the world through the last stages of the Industrial Revolution and into the modern age. Yet if we look at the Brooklyn Bridge and Eiffel Tower, both were constructed during the 1880s and both are riveted together, not welded. In those days, welding processes produced hard, brittle welds that were not suitable for joining large sections of iron or steel under heavy loads. The development of a flux coating on the welding rod was the key to making arc welding more usable, but that took another 50 years to perfect. By the 1930s, welding was being used to construct ships, pipelines, and skyscrapers. For better or worse, war has been the main factor in the advancement of many technologies, including metalworking. The biggest wars—World Wars I and II and the Cold War—precipitated the biggest advances in welding. Research and development of new technologies has trickled down to civilian use.

Modern equipment allows us to easily do welding jobs at home that would take a blacksmith weeks to complete. You can find high-quality steel, aluminum, and other metals in almost any structural shape and size at the local steel yard. Joining them is as easy as pulling the trigger of a wirefeed gun.

Still, welding is a skill difficult to master. Think of it like this: I can put a band-aid on someone's cut and give them an aspirin, but that doesn't make me a doctor. I can change a light fixture and its switch in my home, but that doesn't make me an electrician. Just because someone can lay a bead does not make him or her a welder.

The good news is we no longer have to dedicate our lives to the craft in order to join metals. But we should recognize the difference between what we know how to do and a professional who has spent most of his life learning and working in the trade. New ways of processing, refining, and combining metals, as well as the further development of new welding processes using lasers, electron beams, and friction, indicate that the rate of new discoveries and applications of different metal-joining processes will continue into the foreseeable future. The Internet and books are great resources to learn more about the history of welding and metalworking.

WELDING PROCESSES OVERVIEW

Welding processes can be divided into two basic categories: manual and automated. This book deals with the former, the manual processes, the ones in which the person doing the work has a direct influence on how the weld turns out. This puts the pressure on the person operating the equipment to produce a good weld.

Some of the manual processes are considered semiautomatic, in which one or several aspects of the weld is controlled by the machine. For example, in wirefeed welding the wire is automatically electrified and fed into the molten puddle by the machine. Two processes, which can be manually operated but are more commonly automated, are not covered in this book: plasma arc welding (PAW) and submerged arc welding (SAW). Both have applications in industry, but are not common in most metal fabrication shops because of lack of need or cost of equipment.

Automated is not the best term to describe the wide variety of welding processes in which there is no manual control over the welding process, but the welding does occur in an automatic manner. The only influence an operator has in these cases is to set up the equipment and materials properly.

Some examples of automated process welding are:

- Resistance welding, which includes resistance spot, seam, and projection welding
- Solid-state welding, which includes explosion and friction (stir) welding
- Laser-beam welding and cutting
- Electron-beam welding
- Electroslag welding
- Thermite welding
- Robotic systems (used for welding)

These processes are used extensively in the manufacturing and research industries. The use of robotics and automated processes has grown rapidly in the past few decades and computer numerical controlled (CNC) equipment in machining, welding, and metal forming has revolutionized industry. In today's metalworking shop, the welder may wear many hats, including operating machines that cut, fold, punch, and form metals, so it is important to learn both welding and metal fabrication. If you are working in your own shop you will need to do both out of necessity. The welding processes (and related tools and equipment) covered in this book are common manual processes used every day in metal fabrication shops large and small.

WELD CERTIFICATION

What is a certified welder? There is no simple answer, as many different organizations offer certifications for a wide variety of processes. A certification in welding is different than most other fields due to the variety of possibilities and lack of one central welding union or authority. There are organizations dedicated to the study and advancement of welding, such as the American Welding Society (AWS), which has welding codes and standards used by many organizations to qualify their employees, but there is no one welding test that everyone uses. So, a certified welder is someone who has taken a welder qualification test using guidelines put forth by a company or organization, and performed the test correctly. The welding code or standards also outline how the weld will be tested using destructive, non-destructive, and visual inspection methods and evaluated for pass or fail. Organizations that outline welder qualification requirements include ASME International Boiler and Pressure Vessel Code, Department of Defense Naval Ship Division Code, American Petroleum Institute TD 650 Welded Steel Tanks for Oil Storage, and AWS D1.1 Structural Steel Code, just to name a few.

The code to be used for any one weld certification test will depend upon: the structure or part to be welded (skyscraper, railcar, etc.), the type of material used, the type of filler metal to be used to join the base metals, the material thickness, type of joint design, and in what position the welding is to be done. As each one of these variables changes, so can the test. The possibilities are endless. Typically, even if a welder holds certification papers, the potential employer will qualify him again using the same or different standards to prove that the welder can do the job correctly. For example, a union pipefitter may hold a certification, but will need to take and pass the same test again before going to work retro-fitting a privately held power plant.

Fully automated welding systems, like this robot, are most useful when large quantities of the same part are required. The person programming or running the robot should have a background in welding. Welding education includes information on selecting welding processes, proper machine settings for the material thickness, joint designs, position of welding, and the evaluation of finished welds. A robot still cannot do these on its own. *Monte Swann*

These are plate and pipe welds used in certification. The groove weld on the left has been cut into strips with an oxy/acetylene torch in preparation for a bend test. The pipe on the right is marked with the weld position and welder identification number. The two horseshoe-shaped pieces of metal in the center are the sample pieces from the bend tests. The top is a face bend and the bottom is a side bend sample. *Monte Swann*

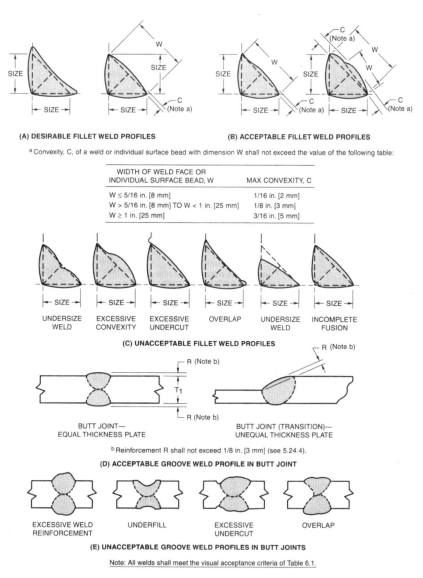

(A) DESIRABLE FILLET WELD PROFILES

(B) ACCEPTABLE FILLET WELD PROFILES

a Convexity, C, of a weld or individual surface bead with dimension W shall not exceed the value of the following table:

WIDTH OF WELD FACE OR INDIVIDUAL SURFACE BEAD, W	MAX CONVEXITY, C
W ≤ 5/16 in. [8 mm]	1/16 in. [2 mm]
W > 5/16 in. [8 mm] TO W < 1 in. [25 mm]	1/8 in. [3 mm]
W ≥ 1 in. [25 mm]	3/16 in. [5 mm]

UNDERSIZE WELD   EXCESSIVE CONVEXITY   EXCESSIVE UNDERCUT   OVERLAP   UNDERSIZE WELD   INCOMPLETE FUSION

(C) UNACCEPTABLE FILLET WELD PROFILES

BUTT JOINT— EQUAL THICKNESS PLATE

BUTT JOINT (TRANSITION)— UNEQUAL THICKNESS PLATE

b Reinforcement R shall not exceed 1/8 in. [3 mm] (see 5.24.4).

(D) ACCEPTABLE GROOVE WELD PROFILE IN BUTT JOINT

EXCESSIVE WELD REINFORCEMENT   UNDERFILL   EXCESSIVE UNDERCUT   OVERLAP

(E) UNACCEPTABLE GROOVE WELD PROFILES IN BUTT JOINTS

Note: All welds shall meet the visual acceptance criteria of Table 6.1.

Figure 5.4—Acceptable and Unacceptable Weld Profiles (see 5.24)

This is a page for the AWS structural steel welding code book. If you take a welded joint and cut it in half, you will be able to see the weld profile. The shape of this profile, which is essentially the shape of the weld bead, is important to the integrity of the joint and the structure. As you can see, certain profiles are unacceptable and are more likely to lead to the weld failing to stay together. *AWS D1.1/D1.1M 2006, Figure 5.4, Reproduced with permission from the American Welding Society (AWS), Miami, FL USA*

Certifications are legal documents attesting to the ability of a person to perform the weld adequately. It can also indicate to potential employers that a welder is familiar with the process and testing procedures. However, just because someone says he is a certified welder doesn't guarantee that he has a wide range of knowledge, or can even weld that well using different types and thicknesses of metals, joint designs, positions, and processes.

You can get certified by joining a union, working in industry, or taking a certification test. Your local AWS chapter or a weld-testing company may provide testing services. The place to start would be a welding program at your local technical college. The welding instructor should be able to answer your questions regarding weld certifications and what the school has to offer. Be aware: taking a certification test usually requires a strong knowledge of welding basics and time practicing and perfecting your skills, with some cost involved.

In this picture, one of the groove weld strips is placed in the vise and the backing plate is marked with soapstone. *Monte Swann*

Next, a fiber cutoff wheel is used to cut most of the way through the backing plate. *Monte Swann*

The backing plate is then removed with a hammer and chisel. *Monte Swann*

The rest of the extra weld metal is ground down and the test strip is carefully sanded flush. *Monte Swann*

7

The sample strip is clamped in a motorized wrap-around-bend test jig. *Monte Swann*

The test strip is wrapped around a 1½-inch diameter mandrel. *Monte Swann*

Once the machine has made a complete cycle, the bent test piece is removed and visually inspected to the criteria of the welding code being used. Two pieces are bend-tested; one is a face bend (top of the weld) and the other a root bend (bottom of the weld). *Monte Swann*

WELDING VS. BRAZING AND SOLDERING

The term "welding" is applied in many ways. Solid-state welding processes produce coalescence, or mixing, of metals at temperatures below the melting point of the base metals, and without the addition of a filler metal. For example, explosion welding, in which the surfaces of two metals become interlocked together in what looks like a series of microscopic waves, is a result of friction caused by the directed force of the explosion. In forge welding, the base metal is heated to white hot, which makes it soft (plasticized) without reaching the melting point, remaining in a solid state as the force of hammering causes the surface molecules to fuse together. There is even a method called cold welding that doesn't require heat, just smooth surfaces and enough pressure to form a bond between them.

Resistance welding is similar to solid-state welding in that time and pressure are key factors; only an electric current is used to generate the heat required for joining the materials. Spot fusion is a common form of resistance welding used to weld ductwork and automobile frames.

In gas and arc welding, the base metals to be joined are heated to their melting point, mixed with a compatible filler metal, then allowed to cool and solidify into one piece. In this kind of welding, coalescence and fusion take place within the joint (not just on the surface) and two separate pieces of metal become one. The depth of penetration is far greater in this case, meaning that the weld zone is much further into the metal's thickness. When the weld has penetrated all the way through the weld joint, this is called 100 percent penetration or complete joint penetration (CJP). Typically the filler metal used when gas and arc welding is similar in content to the base metal. Think of melting ice into water, then mixing in some motor oil and attempting to refreeze the combination. Water and oil don't mix and will not bond. If you did the same experiment with water and orange juice, the water would dilute the juice and the combination would freeze together well. In order for fusion to take place, the materials being welded together must be compatible.

Brazing and soldering are not welding processes at all. These two processes use less heat, far below the base metal's melting point, and a filler metal alloy that is dissimilar to the base metal. The bronze or solder adheres to the surface of the base metal and travels through the joint via capillary action. This method can be a very effective way of joining metals and should be used in certain situations. Refer to the section on brazing and soldering for further information.

THE MOST COMMON WELDING/CUTTING PROCESSES AND THEIR RELATED NAMES

In the welding world there are many names for the same thing. This can be confusing when talking to other people about the subject. Chart 1-1 shows a list of common names used for similar processes. There is a section of the book for each one of these processes.

Communicating with people about welding can be challenging, and few people will like being corrected on

Refer To	Proper Term	Also Known As
Chapter 7	OAW – Oxygen Acetylene Welding	Oxy/Acetylene Gas Welding Oxy/Fuel Welding
Chapter 8	TB - Torch Brazing	Braze Welding Brazing
Chapter 13	OFC – Oxygen Fuel Cutting	Oxy/Acetylene Cutting Torch Cutting
Chapter 10	SMAW – Shielded Metal Arc Welding	Stick Welding Arc Welding (Carbon Arc Cutting)
Chapter 11	GMAW – Gas Metal Arc Welding	Wirefeed Welding (solid wires) MIG – Metal Inert Gas MAG – Metal Arc Gas Short Arc Spray Transfer Welding Pulse Spray Transfer Welding
Chapter 11	FCAW – Flux Core Arc Welding	Wirefeed Welding (tubular wires) FCAW-S (Self Shielded) Inner Shield FCAW-G (Gas Shielded) Dual Shield
Chapter 12	GTAW – Gas Tungsten Arc Welding	TIG – Tungsten Inert Gas Heliarc WIG – Wolfram Inert Gas
Chapter 13	PAC – Plasma Arc Cutting	Plasma Cutting

Chart 1-1
*Colors indicate related gas and arc welding and cutting processes.

their terminology. Be patient and ask questions so you and the steel-yard jockey, welding sales representative, or welding professional are on the same page.

ABOUT THE WELDING EXERCISES

Mild carbon steel is used in all the practice exercises (except one). When you learn how to weld, use this type of material. The steel is relatively cheap compared to stainless steel or aluminum. Mild steel is easiest to weld and is a good choice for practicing your basic welding technique before moving on to the more exotic and expensive materials. Take some time to study the joint designs, types of welds, and welding positions and be prepared by knowing the advantages and limitations of the equipment you are using.

All of the metal I use is cut into 2 × 6-inch pieces. The only exception is Exercise 2 in Chapter 10 (Shielded Metal Arc Welding [SMAW]). When learning to weld, make your beads at least 6 inches long as it's more difficult to maintain good technique on longer beads. These welding exercises are the same ones I use in the classroom and they are meant to be done in the order presented, since the skills learned in each

exercise build off those learned in the previous ones. If you are having trouble with difficult projects, don't be afraid to go back to the earlier exercises and redo them to perfect your technique. The first exercise—running a bead on a plate—is not only about welding properly, but having the machine controls or torch settings right. Make certain your equipment is functioning properly and the correct torch tips, electrodes, filler rods, and shielding gases are used. Having everything running smoothly will allow you to concentrate on improving your welding technique and reduce frustration.

Challenge yourself by making up additional exercises not covered in this book. In my welding classes I typically have students begin by welding all the various joint designs in the flat position. Then we move to horizontal, vertical, and overhead positions. Each new welding position is more difficult than the previous one, and all the combinations of joint designs and welding positions have unique challenges.

The welds in the exercises are to be made over the full length of the material. Remember, these welds are for *practice only*. Later, depending on the project in which you use your welding skills, you may not need to, or want to, make full-length welds.

CHAPTER 2
SAFETY

Safety is a primary concern since there are many ways to get injured or killed when welding. I once asked one of my coworkers to teach me how to use the table saw. Before we started, he asked me, "Are you afraid of it?" The saw wasn't even turned on and I said yes. He said "Good, because as soon as you aren't afraid of it you will lose a finger." That made sense to me. As soon as you lose respect for something dangerous it is liable to bite you.

The owner's manual that comes with your welding/cutting and fabricating equipment is full of good and useful information. Read through it; know how your equipment works and the specific hazards related to its use.

Take one minute to consider the consequences of your actions. This is especially true when cutting. Steel is strong and supports heavy weight very well, but can be cut easily with a torch. A drum, cylinder, or container can also be cut with a torch, but without knowing exactly what is inside it you could be in for a lot of trouble. Leaning into a belt sander or pushing metal through a vertical band saw requires only one slip before disaster. Spinning pulleys, flywheels, and drill chucks are hazards. If your hair or clothing get wrapped around a motorized turning part or caught in a pinch point where moving and non-moving parts come together, it will pull you in. And in the case of someone's hair, the machine can actually scalp you.

Whenever you start any machine, ensure that you are not in the path of any shrapnel should something come apart. Stand to the side of any machine, like a mounted bench grinder. Or for example, after installing a new hard wheel, hold a 4½-inch angle grinder at arms length with the spindle in a horizontal position pointing away from you. Think about what direction things are moving, or could move in, and avoid that area. Don't forget: something doesn't have to be moving in order to inflict severe injury. You can run into sharp metal or machinery. Running and horseplay in a metal shop is asking for a laceration or to be impaled on anything laying around. Alcohol and drugs are also a bad idea in a metal fabrication shop. If you shouldn't be driving, you shouldn't be welding.

SHOP SAFETY PRACTICES
Setting up shop
Before you begin to buy equipment, the first thing to consider is the location of your shop. Welding and cutting in a basement is not a good idea since welding smoke is lighter than air and will travel to the floors above. A detached garage or other building is ideal.

The area where welding and cutting is taking place should be well ventilated. Most people solve this problem by opening the garage door or working outside. This is a good solution if you are using welding processes that don't require use of a shielding gas such as SMAW, FCAW-S, and oxy/acetylene. The processes that do use a shielding gas, such as GMAW, FCAW-G, and GTAW, are wind sensitive. Even the air movement generated by a fan can blow away the shielding gas, contaminating your weld and ruining it. Of these wind sensitive processes, some generate more fumes than others. For example, GTAW generates the least amount of fumes, GMAW next, and FCAW-G generates the most.

Your decision to install an exhaust system or use a respirator will depend upon which welding and cutting processes you are performing and with what type of materials you will be working. Also consider that when you heat certain materials to their melting point, you send very small particles into the atmosphere. Breathing in enough of these particles will make you sick in the short term and increase your risk of long-term health problems. See the section on respirators in this chapter for more information. Some simple ductwork and an exhaust fan can go a long way in reducing the amount of exposure to toxic fumes, but may not be the most practical or useful solution. If you decide to install ventilation, go to the OSHA website (www.osha.gov) and search for OSHA standard 1910.252(c)(3)(i). The standard gives you duct diameters and minimum airflow rates in relation to the distance the vent is from the weld zone. Your vent system

Notice the welding fumes, which are lighter than air, float up from the weld and get sucked into a vent positioned above the weld joint. Welding in the vertical position as shown will keep your head out of the smoke plume. When welding in the flat and horizontal positions, be aware of keeping your head out of the plume. *Monte Swann*

Along with an ABC fire extinguisher and first aid kit, a fire blanket can be used to snuff out fires. In addition, hot pieces of metal can be wrapped in the blanket to slow the cooling rate of certain metals after welding. Because it is located at a public school, this fire extinguisher is tagged with dates of inspection. The directions on this fire extinguisher explain its proper use. When you need to use an extinguisher, remember PASS: Pull the pin, Aim for the base of the fire, Squeeze the handle, and Sweep back and forth. *Monte Swann*

can be ambient capture, meaning an overhead hood or down-draft table, or source capture, consisting of a moveable arm. A moveable arm provides the most versatility, because it can be moved closer to the source of any smoke and fumes.

Keep safety in mind

Working with electric arcs, flames, and molten metal greatly increases the risk of starting a fire. Before you start welding or cutting, be certain to check the work area for any flammable materials, like rags and cardboard, and remove them. Flammable liquids—such as oil-based paint, finishes, solvents, and gasoline—even if in a metal container, should also be removed from the area or stored in a flammable materials cabinet. Batteries pose an explosion hazard; never weld near batteries. Buy a couple of ABC fire extinguishers and have them handy. The very small extinguishers will not last long when it counts the most. I recommend purchasing a 6-7-pound extinguisher, at a minimum. Don't be afraid to buy a larger one.

A metal bucket full of water is good to have around for cooling metal after practice welding. Never use water to extinguish a fire around molten metal, because water will react violently with the molten metal. A bucket of sand helps put out any flames. Different sizes of scrap sheet metal come in very handy as a barrier between your welding or cutting and something hazardous. For example, if sparks gather in openings between and underneath walls or come in contact with wood floors, they can smolder and possibly catch on fire. Concrete floors and regular bricks have moisture trapped inside. When a hot flame from a torch comes close enough, it will turn the water into steam, which will expand

rapidly causing an explosion, sending concrete fragments in all directions at a high rate of speed. Automotive work is especially hazardous due to fuel lines, fuel tanks, and flammable upholstery. Use sheet-metal barriers to deflect heat and sparks away from dangerous areas.

A bottle of eyewash solution is handy for flushing unwanted fragments or liquids out of your eyes, and could help in an emergency. Let someone know where you are and what you are doing if you are working solo. Have a well-stocked first aid kit with tweezers, burn cream, disinfectant, bandages, and tape to take care of the minor injuries you will face in the shop. Finally, make it a policy to hang around the shop for one-half hour after all welding and cutting is finished. It will give you peace of mind to know that your shop is not on fire after you leave.

Keeping a (relatively) clean and well-organized work area will help in everything you do. It is a lot easier to trip and fall over debris, tangled welding hoses, or cables in a messy shop, and anything you fall into will likely go right through your clothing and skin. Cables and cords usually need to be on the floor, so have out only what you need to do the current job. Keep the rest wrapped up and be careful walking around. Sweep after you finish working since steel dust is slippery on some floor surfaces.

BUYING WELDING EQUIPMENT

Deciding what type of welding machine and equipment to buy can be difficult. There are some basic questions you need to answer before purchasing anything. What type of materials will you be working with? Steel only, aluminum and/or stainless steel, as well? Some welding processes, such as GTAW, are well suited to welding a variety of materials. Others, such as SMAW, would be a bad choice for welding aluminum. There is more information on the advantages and disadvantages of each process later in the book.

The next thing to consider is the thickness of the metals to be welded and, more importantly, the maximum material thickness you want to weld. The capacity your machine has is directly related to how much heat it can generate. You need more heat to weld thicker sections and less to weld sheet metal. The capacity or amperage of a machine is also directly related to cost. With some processes, such as SMAW, you can weld thicker sections for a lower cost than if you were using GMAW. Processes like GMAW will be significantly more expensive to weld 3/8-inch material and over, but require less skill. This will be discussed later in the book.

Where will you be welding? Indoors, outdoors, in a garage with the doors open? As we discussed earlier, this is an important consideration because some processes are wind sensitive. Whatever machine or torch setup you decide to buy, I strongly suggest going to your local welding supplier. You can find one near you in the phone book and it will be well worth the effort to visit one or two. It will help if you read up on welding a little and have some ideas going in, but the people working there will be much more knowledgeable about the subject than employees of big chain hardware or home stores. Stores that have the welding machines next to table saws and drywall have a limited variety of equipment to choose from and often carry machines that will not ultimately suit your needs. A good welding supplier will take the time to answer questions, as well as point you in the direction of some other useful accessories. Decide on a budget, but don't go cheap. How much you pay for a welding machine will depend upon what process you will be using. But a broad range of prices for a good welding machine range between $600 and $3500. Buying a machine up-front that will do the jobs you want to do should offset the cost of buying a cheap machine first, using it for a couple of years and then buying another, better machine. See Chapter 9 (Introduction to Arc Welding) for more information on welding machines.

When you buy your machine, be certain to stock up on all the consumables you will need. Parts of a wirefeed gun, for example, are considered consumables, and some are replaced less frequently than others. But having extra contact tips and nozzles will be helpful during a fabrication project. Having extra sanding/grinding wheels, filler rods, tips, torch, and gun parts on hand can be the difference between finishing a project and waiting. If you are lucky enough to inherit welding equipment from a family member or friend, find out exactly what you are getting into. What will the equipment do and not do? Is it old and in need of a tune-up? This is where a welding supplier can come in handy, setting you up with new welding cables, a new torch or wirefeed gun, gas lines, bench-testing regulators, and tuning up your machine. The cost to do these things is reasonable and could save you a lot of trouble down the road. Used equipment can be great, but could cause a lot of frustration if it's acting funny or be a pain if a regulator blows or there is a fire or you get shocked due to an electrical problem.

No matter what size space you will be working in, portable equipment is wonderful. I put as many things on wheels as possible. Welding machines, tables, and material racks can all be made mobile by welding up a simple frame out of angle iron or steel tubing, and then welding or bolting casters to the bottom. I had a student who modified a two-wheeled cart to fit his smaller GTAW machine and shielding gas cylinder. He mounted the machine vertically with the controls facing up, and the back of the machine was mounted away from the floor to accommodate the cooling fan. The other side had the cylinder chained in the normal position. With today's lightweight machines, there are many more possibilities. Putting things on wheels gives you greater versatility and access in your shop. It's a lifesaver if you need room in the garage for your car, too. Once you have some welding equipment, materials, and skill you can fabricate all kinds of brackets, shelving, and storage, as well as modify any store-bought items. Custom make what will work best for you. Have a home for tools so the next time they can be found easily. And always keep safety in mind.

PERSONAL PROTECTIVE EQUIPMENT (PPE)

Personal protective equipment is just a fancy way of saying what you should wear in the shop. What you'll need depends on the particular job you are doing. During my first semester of welding school we used track torches to cut bevels on thick steel plates, which is a gas cutting-torch on a motorized track. As I was in the middle of watching the track torch do its work with my dark cutting goggles on, one of the instructors walked behind me and said casually, "Your shoe is on fire," and kept on walking. I took off the goggles and looked down to see my canvas shoe in flames. After I put the fire out, I went to my locker and pulled out the leather work boots I had chosen not to wear. I was not dressed properly for the job at hand and got lucky not to be severely burned. Working with the hazards of welding requires proper dress, and even then your clothing can catch on fire. Keep in mind, the risk of being burned increases depending on the welding process, amount of heat and sparks, and how well you are protected by what you are wearing. Wearing the proper protective equipment for the job will greatly reduce the risk of personal injury, especially if the work involves cutting, grinding, and welding out of position (vertical and overhead).

What you should not wear

Nylon and polyester clothing should not be worn. Heat from welding or cutting can cause the material to melt to your skin. The unfortunate people who have had this happen need skin grafts to replace the damaged area on their bodies. Wrist-watches, rings, and jewelry should be removed because they heat up quickly and can get caught on materials, shelves, and moving parts. Long hair should be tied back. Sandals and open-toed footwear are a bad idea.

Upper and lower body

A good pair of jeans is all I've ever worn in the shop, at least between my ankles and waist. The jeans you wear should be the correct length and not be cuffed at the bottom. A cuff on your jeans or shirt, or even a shirt pocket without a cover flap, is a spark collector. Sparks like to be with other sparks and if enough of them end up in your cuffs, there will be a fire. Your jeans should be in good condition without any frays. At a minimum, put some tape over the frayed area, but only as a temporary measure. What you wear on top depends on what kind of work you are doing.

Heavy-cotton, long-sleeve work shirt made of denim (not a fuzzy nap, like a sweat shirt)
- Gas welding and brazing in all positions, TIG (GTAW) welding in all positions
- Light grinding, torch cutting, and plasma cutting

Welding jacket made from green (typically) fire-retardant material
- Light wirefeed welding (GMAW/FCAW wires) in all positions

- Medium wirefeed welding (GMAW/FCAW wires) in the flat and horizontal positions
- Light to medium stick-welding (SMAW) in the flat and horizontal positions
- Medium torch and plasma cutting and heavy grinding

Welding leathers or leather jacket
- Medium to heavy wirefeed welding (GMAW/FCAW) in all positions
- Medium to heavy stick-welding (SMAW) in all positions
- Heavy torch and plasma cutting

This is an all-cotton welding jacket. The material has been treated to be fire-retardant. These jackets are inexpensive. You can also buy welding sleeves made out of the same material to be worn over your bare arms or shirt. *Monte Swann*

This welding jacket is known as a hybrid. The arms are made of leather and the body of the jacket is fire-retardant cotton material. Since your arms are closest to the sparks and liquid metal, they need a high level of protection offered by the leather. *Monte Swann*

This is a pair of welding leathers with an apron snapped onto the front. An all-leather welding jacket is also available for welding with high levels of heat and sparks and/or welding vertical and overhead joints. *Monte Swann*

Light, medium, and heavy are terms used above to describe the amount of heat, sparks, and intensity of arc light generated by the welding or cutting process and the direction it's traveling. For example, in overhead welding you will likely be under the weld where sparks, molten metal, and hot slag can drop down—a much more hazardous position than having the weld on a table in front of you. It is not a bad idea to have each level of protective clothing available, so if you do get burned you can suit up better for the job. I avoid wearing leather aprons because they can get caught in moving machinery, but having a piece of leather available can be a great way to protect you from heat and sparks in certain situations. For example, a piece of leather may be placed in your lap if you are welding while sitting on the floor.

Gloves

Another reason to visit a welding supply store is the wide variety of proper gloves. Welding gloves are all leather with a full leather gauntlet that covers the end of your sleeve. They range in heat protection from oven-mitts to almost paper thin. The sacrifice you make with wearing heavy gloves is a great reduction in manual dexterity. One wrong move with thinner gloves and they are easily burned up. Again, it depends on what kind of welding you are doing. I wear a pair of Tillman No. 30 TIG gloves for gas and TIG welding. I also wear a heavy cotton shirt. The more heat protection required up top, the thicker the gloves should be. I suggest buying one pair for lower heat and one for high heat. Your welding supplier will have a wide variety of welding gloves for all welding applications and can help you find the ones right for your job. Tillman and Revco are two good brands to look for. Revco also makes a women's line of welding gear. Keep your welding gloves dry and in good condition. Never pick up hot metal with them or they will shrivel up and be ruined quickly. Always use pliers when picking up hot metal. Keep a pair of old gloves for material handling. Oftentimes, steel you get from a yard has oil, grime, and dirt on it. Oily gloves and welding equipment are a bad combination.

It's handy (no pun intended) to have different welding gloves available for use in the shop. Watch out for holes developing in the gloves during extended use. Gloves with holes can be patched with duct tape and are good for handling materials with sharp edges. *Monte Swann*

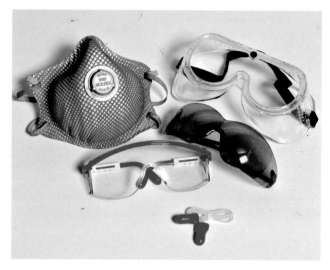

The red-framed safety glasses have a standard, clear polycarbonate lens. The darker-framed glasses have shaded lenses for working outdoors on sunny days. The goggles at the top of the picture can fit over regular prescription glasses and protect your eyes from flying particles. *Monte Swann*

Shoes

All-leather, steel-toe, high-top boots offer the most protection. Leather shoes also work well, but it takes only one heavy object falling on your foot to convince you steel toes are a good thing. Steel toes have another advantage when it comes to moving materials around or kicking something hard. Often I find myself using the end of my steel toes as a brace or extra hand. Keep in mind, a very heavy plate of steel or fork lift will crush the steel in your shoe, but this is an unusual occurrence, even in an industrial shop. Unless you wear steel-toe slip-on boots, extra shoelaces come in handy in case the pair on your shoes gets burned up.

Safety glasses

I ask my students at the beginning of the term, "If you had to choose between losing a finger or losing an eye, which would you choose?' Always, they choose the finger. Then I up the ante and ask, how about two fingers? Sometimes we get to the whole hand before someone thinks twice. No one wants to lose an eye. Always wear your safety glasses in a shop. They are one of the least expensive things to buy and protect one of the most important parts of your body. Buy a clear comfortable pair that resists fogging. Keep them in good condition; scratches and pits can obstruct your view of the molten weld pool, making it more difficult to see what is happening. If you already wear glasses, they might be okay. The lenses should be made of polycarbonate and large enough to cover your eye socket. They should also have side shields. Specialty safety glasses referred to as flash goggles are a No. 2 or 3 shade. The shade reduces the brightness and annoyance of other welding arcs.

Contact lenses are okay to wear while welding, and no matter what you have heard, they will not melt to your eyeball. The only concern is that they can dry out quickly, so keep them moisturized. Seeing the weld is very important. Many times people learning to weld have a hard time simply because they cannot see clearly. Bifocals under your welding helmet can help, and the magnification can be placed on the lower or upper edge of the lens, or both. Instead of bifocals, you can ask your welding supplier to show you their selection of cheater lenses. These are magnifiers that fit into your welding helmet and come in a variety of strengths. In any case, you should be able to read small print up close, about 6 inches from your nose, as well as at arms length.

Ear protection

Metal fabrication can be a very loud business. Grinding, cutting, and hammering will put an extreme amount of stress on your ears. Protect your hearing by having a box of ear plugs and wearing them when the noise level gets uncomfortable. I like to use ones with a cord attached between the plugs making them easy to insert and remove. Hearing loss is compounded over time, followed by a constant ringing in your ears. Like losing an eye, the effects will last the rest of your life.

Face shields are available with a clear lens or a No. 5 shade. The same headgear can be used for both. *Monte Swann*

A clear face shield and gloves are a good choice when using grinders. Notice the sparks from a grinder come off the wheel in one direction. Be sure to keep your spark stream away from flammable materials, compressed gas cylinders, welding equipment, and other people. *Monte Swann*

Face shield

Consider investing in two face shields: a clear shield and a No. 5 shade. The clear face shield is worn when using an abrasive cutoff wheel and grinding with a hard-wheel attached to a rotating tool. Both of these wheels are very useful, but can break apart under certain circumstances. If they come apart, pieces of or the entire wheel becomes high-speed shrapnel. The No. 5 shade is used with oxy/acetylene equipment for gas welding, brazing, cutting, and heating. It is more comfortable than wearing cutting goggles and protects you from flying sparks and molten metal.

Welding helmet

There are two kinds of welding helmets available today. In the old days, the only type of helmet was one that was flipped down just before welding. As a welding student, one of the first skills learned was keeping your hands in the same position while nodding your head to bring the helmet down. The moment the helmet is down everything is dark, because the shade lens required to look at the welding arc is too dark to see through otherwise. Keeping your hands steady and striking the arc at the beginning of the weld when you can't see where it is, is crucial.

The auto-darkening helmet has changed that. Now everything can be seen before welding begins, which is a great help for the beginner, as well as making the professional more efficient at his job by not having to lift up and flip down the helmet as often. However, auto-darkening helmets cost more and can break or become damaged permanently. Regular helmets with the proper shade lens work great, cost little, and do not break in a way that can't be fixed. I suggest buying one with a large window; Jackson makes a good one with a 4½ × 5¼-inch viewing area which will help you find

The interior of this auto-darkening helmet shows all the settings available for a wide variety of welding situations. Any weld spatter or hot sparks can damage the interior side of the lens. Be certain to use clear protective plastic pieces to prevent ruining the helmet, and get in the habit of laying your helmet down with the face up when you are not using it. *Monte Swann*

With the front cover off, we can see the small square sensors on each side of the viewing window. The solar panel below powers this auto-darkening helmet while the arc is on and batteries power it when the arc is off. *Monte Swann*

SAFETY

Auto darkening helmets are great, but I would also have a standard helmet as a backup. The No.10 shade lens that comes with most standard helmets is cheap and usually made of plastic, which is easily scratched. For a little more money it's worth buying a set of better-quality shaded lenses made of glass in a No. 9, 10, 11, and 12. *Monte Swann*

the arc easily. However, large windows reflect more of what is behind you. Other people prefer a smaller window because there is less area to look at and, therefore, fewer distractions. On many helmets with smaller windows, the shaded lens flips up so the helmet can be used like a face shield for chipping slag and using a power brush.

There are a few considerations when buying an auto-darkening welding helmet. Look for one with multiple shade settings, from shade 9 to 11 or 13. You may need to use different shades to see well. Make certain the helmet has a sensitivity switch or a delay, especially if you plan on doing any TIG welding. If you use a regular welding helmet, get a variety of shades, as well, from No. 9 to 11. Regardless of the helmet type used, you need extra sets of clear lenses, which

Both regular shaded lenses and auto-darkening lenses are protected by clear plastic inserts. Have a few of these inserts on hand so you can change them out when they get dirty, pitted, and scratched up. Being able to see your work clearly is the first critical step to being a successful welder. I'm holding a cheater lens, which magnifies what you see much like reading glasses or bifocals. These are a big help for people without perfect nearsighted vision. *Monte Swann*

are protective covers for the shade components underneath. There is a range of prices and quality available. Plan to spend between $150 to $325 for a good auto-darkening helmet or lens. Think twice before trying to save a few bucks on your eye protection.

Skull cap

Welding suppliers have skull caps for free or at a small cost. If the company advertises the business name on the skull cap, it should be free. The skull cap protects the top of your head from sparks and helps mop up sweat on your brow. A baseball hat turned backwards or a bandana tied around your head will do the same job.

FUMES

The respiratory hazards associated with welding come in two forms: particles and vapors. These can come from welding, brazing, and cutting operations involving heat, grinding, and polishing. Welding fumes, which consist of particles, not vapors, form from evaporation of liquefied metal droplets, spatter, and the weld pool. These particles clump together forming particle clusters one micron in size. The good news is welding particles are easier to filter than most other particles of the same small size. A half-mask respirator will go a long way in attracting those particles into the fibers of the mask, before you breathe them into your lungs. A half-mask respirator is not the same as a dust mask. Dust masks have only one strap, respirators have two. If you get the disposable kind it should have an N rating of 95, 99, or 100. The N stands for non-oil resistant and is fine for welding. R- and P-rated filters are oil resistant and are used mostly in machine shops where

spraying cutting fluids are present. Look for this number on the packaging or on the mask's one-way exhaust vent. If you want to spend a little more, buy the non-disposable half face piece with replaceable cartridges that protect you from particles and/or toxic vapors. Be certain to have and install the correct cartridge for what you will be exposed to. Keep in mind that the respirator has to fit under your helmet if you are using it for welding. Whichever type you use, make certain it fits properly and creates a seal around your nose and mouth. Facial hair and even stubble will greatly reduce a respirator's effectiveness.

The following factors affect the amount of exposure you may experience.

• *Welding position*
  Welding fumes are lighter than air, so they will travel in a predictable direction. When you are welding out of position, vertical, and overhead, most of the time your face will be out of the path of the fumes. Welding becomes much more difficult in these positions, so most welding is done in the flat and horizontal positions. Keep your head out of the smoke plume when welding something flat on a table or if your head is above what is being welded.

• *Welding process*
  Stick welding (SMAW) and welding with a cored wire (FCAW) both involve the use of a flux and generate a larger amount of particles. High voltages and amperages generate more heat for welding thicker materials and generate more welding particles due to the increased arc and weld pool sizes.

• *Base metals*
  Galvanized steel has a rust-resistant zinc coating that creates a harmful vapor when burned off. Overexposure to zinc, cadmium, copper, or magnesium will make you very sick. You might feel it right away or later that evening. The symptoms are flu-like, including nausea, sweating, dizziness, sore throat, and fatigue. Symptoms will subside after a day or two, but those who have had metal fume fever or zinc poisoning will tell you to avoid it in the first place. Drinking lots of milk will help reduce some of the symptoms, but will not prevent or cure metal fume fever.

  Paint, powder coating, grease, oil, epoxies, glue, and rubber will burn off around the weld and cutting zones creating a toxic vapor. For example, chlorinated solvent vapors and refrigerants create phosgene gas, which is similar to mustard gas used during World War I. Of course, this is extremely hazardous and should be avoided. Protect yourself from these byproducts before repairing or modifying parts and equipment. If the risk is too great, it is best to avoid the work altogether.

Stainless steel contains heavy metals that give it corrosion resistance. Nickel and chromium are most common and are known carcinogens. Hexavalent chromium, which is 47–62 percent of the total chromium fume in FCAW and SMAW, and only 4 percent in GMAW and GTAW, is particularly harmful. OSHA has lowered the permissible exposure limit (PEL) of hexavalent chromium, causing increased protection requirements for welders working with stainless steels. Prolonged exposure could result in major health problems down the road.

- *Manganese*
  Manganese is used in the fluxes and coatings in the SMAW and FCAW welding processes. Overexposure can affect the central nervous system, resulting in impaired speech and movement. Once poisoned by this highly toxic material, the effects are irreversible.

- *Cadmium*
  Used in some brazing filler metals or for plating on bolts and hardware, cadmium is a heavy metal. When ingested, cadmium can cause metal fume fever and is a known carcinogen.

- *Lung cancer*
  Studies estimate that there is about a 30–40 percent increased risk of lung cancer in people who weld for a living. Much of this is due to a high rate of cigarette smoking among welders and exposure to asbestos, which used to be an ingredient in the flux coatings on electrodes. Prolonged exposure to stainless-steel welding is suspected as a link, but is difficult to prove because most welders have joined a wide variety of materials over their careers. Conclusive evidence is still needed to estimate the risks properly.

- *Shielding and fuel gases*
  Be aware of the hazards if you are using compressed gases in welding. Once the gases displace the atmosphere you are breathing, and there is no more oxygen left, you will lose consciousness and die. In this situation a half-mask respirator of any kind will not protect you from fatal consequences. Check for leaks in the system and keep track of the gases you are using. If you get a headache or are dizzy or nauseous, get away from the area and into some fresh air.

CONFINED SPACES

If you ever find yourself crawling or climbing into a confined space to do some work, take great caution before going in. The breathable atmosphere is limited, and once displaced by fumes or vapors, death will result in a short amount of time. This has happened to many unfortunate people, and as a result there are many OSHA regulations concerning working in confined spaces. At a minimum you should have a respirator with supplied air and a spotter, someone who can rescue you in an emergency.

MSDS

That little piece of paper you sometimes get with welding rods, wire, and flux is packed with good information. MSDS stands for material safety data sheet and by law must be provided to those using hazardous products. Every MSDS has sections on general hazards, how the product reacts with other substances (reactivity), and how to handle and use the product safely. The first section of every MSDS identifies the product and gives the contact information of the manufacturer. The second section is a list of ingredients and the threshold limit value (TLV)—how much of the stuff can be in the air (by volume) before creating a hazardous environment. Also in this section are the PELs, which give the maximum amount of each hazardous ingredient (again by volume) that can be ingested before adverse health effects occur. The lower the PEL, the more toxic the substance.

CHAPTER 3
GENERAL TOOLS AND EQUIPMENT FOR METAL FABRICATION

Tools are an important part of any shop. If you take a class at a well-equipped welding lab, you will have access to industrial-grade equipment for fabricating metal parts including shears, ironworkers, band saws, press brakes, and maybe even computer-controlled plasma cutters and robots. Most people don't have the budget or the need for a robotic welding system, but there are a few tools you are going to need. The equipment in your shop will depend on what types and thicknesses of metal you will be working with, the types of projects you plan on building, and the welding equipment you'll be using.

Small hand tools for fabricating metal parts can be powered by electricity (plug in), batteries (portable), or compressed air (pneumatic). An air compressor is necessary for plasma cutting and carbon arc cutting (covered later) so you may already have that piece of equipment, making it reasonable to buy other pneumatic tools. If you don't have a need for an air compressor, then the equivalent type of electrical tool can be used. Have two or three extension cords available; often it is easier to bring the tool to the work instead of bringing the work to the tool.

Thanks to lithium ion battery technology, portable tools have become more powerful and practical to use. It is annoying when they run out of power in the middle of a job, but changing out the batteries is quick, as long as you have another one charged and ready to go.

I've broken up this chapter into six main sections, with a list of commonly used tools in metal fabrication. Some have specific uses, advantages, and disadvantages. No one tool does it all, but some tools and equipment will serve your needs better than others. Be certain to read, understand, and follow all the instructions that come with power tools. Knowing how to use them safely and effectively will reduce your risk of getting injured.

HAND TOOLS
Clamps
Wood workers say you can never have enough clamps. Metal shops are in a similar situation, except we can use tack welds to hold things in place and improvise clamps with fabricated structural shapes. When you do need a clamp it is nice to have them. I would recommend having at least four C-clamps; the quick adjusting kind are expensive, costing between $100 to $180 each, but can save lots of time. Standard C-clamps will take longer to adjust but will cost only $20 to $70. The clamp size will depend upon the thickness of your worktable and size of the components for your project. Keep old or damaged C-clamps for a backup or when the need arises to modify them for certain jobs. Bar clamps are sold with two end pieces that you attach to the correct-diameter pipe. The convenient part is that if you need a longer bar clamp, you use a longer piece of pipe. Be aware there is a limit as to how long the pipe can be without flexing as pressure is applied at both ends.

A vise is a handy tool in any metal fabrication shop. Attach a sturdy, heavy-duty vise to your worktable to hold metal while heating, cutting, grinding, bending, or welding. *Monte Swann*

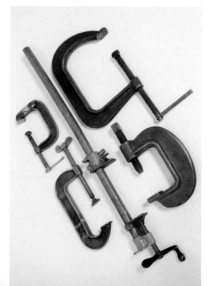

This bar clamp is flanked by four C-clamps. Notice the two smaller C-clamps have been repaired by careful welding and the addition of reinforcement pieces. Broken or cut C-clamps are an opportunity to practice your skills and make something useful again. *Monte Swann*

Pliers

There are so many uses for this tool, but in welding the most important is the handling of hot metal. Use pliers instead of gloves to pick up metal that has just been welded or to hold small pieces of metal in place when welding. A pair of needle-nose pliers with a wire cutter is very useful for cutting down long filler rods or excess wire on the end of your wirefeed gun and pushing down on piece of sheet metal when tacking to close any gaps between the two pieces. Welpers are a specialty tool designed specifically for welding work. This tool is especially useful when wirefeed welding. Tongs are wonderful to have when using a large amount of heat on a piece of metal. Buying tongs new can be expensive, but you can make your own tongs fairly easily in a well-equipped shop.

Vise grips

You can never have enough vise grips in your shop. Vise grips come in all shapes and sizes. There are C-styles for clamping materials together or down to a fixture or work surface. Some are designed specifically for sheet-metal work. Vise grips designed for general use come in handy for holding medium to large pieces of metal when grinding or sanding. They're also useful for holding small pieces of metal, putting distance between your hands and the abrasives, so if you slip, the vise grips get sanded or ground down instead of your fingers.

Wire brushes

Keep a large wire brush around for removing slag from welds and rust from unwanted areas like the weld joint. Steel brushes can be used to clean steel, but can transfer unwanted carbon to materials like aluminum and stainless steel. Wire brushes made from stainless steel are good for cleaning steel, stainless steel, and aluminum, without risking transferring carbon to the base metals. Sandpaper, Scotch-Brite pads, and even wire wheels mounted to an angle grinder are useful for preparing your base metal for welding.

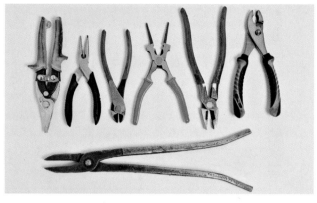

These are some of the hand tools you may want in your shop. The long-handled tongs at the bottom of this photo keep hands a longer distance from sparks, hot metal, and quenching fluids such as water, brine, and oil. The row of tools from left to right are tin snips, needle-nose pliers with wire cutters, heavier duty cutoff wire cutters, welpers, lineman's pliers used by electricians, and regular slip-joint pliers. *Monte Swann*

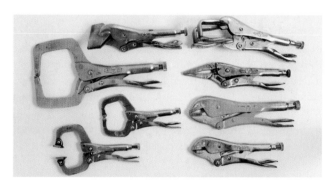

A variety of vise grips will come in handy. Notice the C-shaped Vise-Grip in the lower left corner has metal pads to prevent marring or denting softer metals like aluminum and brass or thinner pieces of metal. Strips of brass or aluminum can also be put between the material and Vise-Grips, or the jaws of a vise to prevent damaging sensitive surfaces. *Monte Swann*

Don't forget to have sandpaper, stainless-steel brushes, and abrasive pads to clean your metal parts before welding. Paint, rust, scale, dirt, or oil left on the metal during welding can greatly weaken its strength and toughness. *Monte Swann*

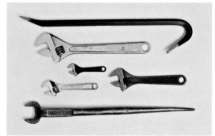

A pry bar is pictured at the top, and adjustable wrenches of different sizes are in the middle. Both of these kinds of tools have many uses in a metal shop. At the bottom is a spud wrench used by ironworkers. The tapered end is used to line up bolt or rivet holes on beams. *Monte Swann*

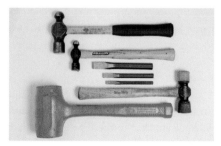

The big orange hammer is called a dead-blow hammer. Loose pellets inside the head prevent the hammer from bouncing back off the metal. Soft hammers are handy when persuading pieces to move without marring the surface. A set of chisels are pictured in the middle and two ball-peen hammers, 8- and 16-ounce, are at the top. *Monte Swann*

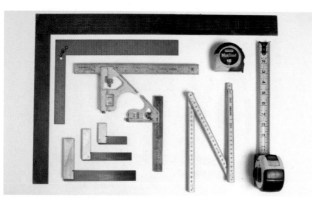

On the left top are two framing squares, two combination squares are in the middle, and a set of machinist squares are on the bottom left. Use squares and tape measures for layout work and in fabrication to confirm that parts are the right length and oriented correctly. *Monte Swann*

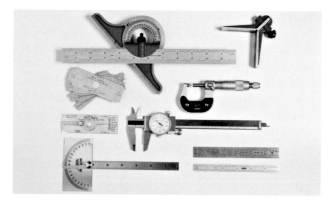

At the top are accessories in a combination square set. In the center are a micrometer, dial caliper, and gauges to measure welds and weld profiles. A protractor and 6-inch machinist rulers are along the bottom. *Monte Swann*

MEASURING TOOLS

Tape measures

A must-have for any project, tape measures come in a variety of lengths. You will find uses for a shorter tape measure (8-foot), which is easier to handle when making short, exact measurements and bends easily around contours like pipe circumferences. The steel strapping used to make long tape measures (25-foot) is stiffer and wider, making them useful for measuring larger projects. Tape measures made of cloth also work well but can be damaged and burned up easily.

Squares

Combination squares are frequently used to measure and mark metal and to check if pieces are square. Many have a built-in small leveling bubble. Combination squares combine the functions of a rule, square, and protractor, and they include a V-shape piece designed for finding the center on round pieces of metal. The ruler in these combination squares are typically made of metal and can easily be ruined by spatter, hot metal, and welding arcs.

Framing squares are useful to have around and can be found in any hardware store, but tend to be the least accurate. Machinist squares are very accurate and are frequently used by welders.

Calipers

A caliper is the best tool to quickly and accurately check sheet thickness, the wall thickness of tubing, and the diameter of pipe. Making accurate measurements is important when purchasing metal at a steel yard and setting up your equipment for welding. A 12-inch dial caliper is easy to read and will take measurements as small as 0.001 inch (one-thousandth of an inch). In manufacturing, decimal lengths are given three places from the decimal point and expressed in thousandths. Half-inch-thick material would be 0.500 inch, or five-hundred-thousandths. Sheet metal 0.020 inch would be twenty-thousandths. A fraction to decimal equivalent chart is useful in making quick conversions. Otherwise, divide the top number of a fraction by the bottom number to convert it into decimal form. As with combination squares and micrometers, take good care of these measuring instruments. They can become worthless if mishandled, neglected, or misused.

Micrometers

Micrometers are precise down to 0.0001 inch. That level of precision is invaluable in machine shops, but not generally needed or useful when welding. If you plan to have any need to measure machined parts it is a good idea to learn how to read a micrometer. Digital calipers and micrometers are available, making them easy to read, but are one hot spark away from becoming scrap metal.

Levels and lasers

Lasers have become common tools used in many industries. A magnetic torpedo level with a laser is a very useful tool when making larger projects. Keeping large frames square and parallel can require a lot of skill. Levels and lasers will help in making accurate measurements.

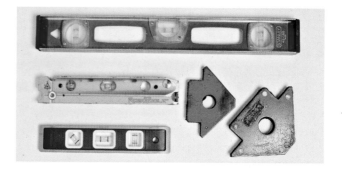

On the bottom are two standard levels with a laser level in the center. The magnetic squares on the right are used to hold steel in place during fit-up and welding. Unfortunately, steel dust and spatter can collect on the magnetic edges and are a nuisance. New model magnets have an on-off switch to solve this problem. *Monte Swann*

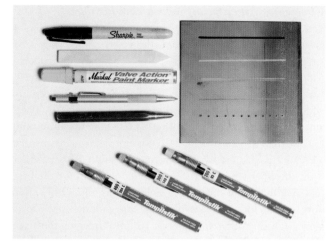

The mark each different tool or marker makes can be seen on the piece of hot-rolled steel on the right. The three red wax crayons on the bottom are not for marking metal. These are temp-sticks used to measure the temperature of the base metal before and during welding. Each type of wax has a different melting temperature. If it smears when applied to a piece of metal, that area of the metal is at a temperature higher than the one marked on the temp-stick. If the crayon does not smear, that part of the metal is at a lower temperature. *Monte Swann*

MARKING TOOLS
Scribes

There are two varieties of scribes: carbide-tipped and diamond-tipped. Carbide-tipped scribers are more commonly used in shops and will mark most surfaces accurately. Keep in mind a scribe marks the metal by creating a groove in the metal, sometimes creating a burr. If your finished piece needs to look good, a scribe can wreck its cosmetic appearance. In machine shops, a liquid known as bluing is used to coat the surface of a piece of metal. Dry, light pressure from a scribe will scratch through the bluing, revealing the metal beneath and making a highly visible mark. Keep in mind bluing is flammable.

Soapstone

Both round and rectangular sticks of soapstone can be used to make non-permanent marks on metal, and they can be sharpened to a point with a sanding abrasive for more accuracy. The advantage of doing layout work with soapstone is that the mark will not disappear when the metal is heated. This is especially useful in torch cutting, where the metal can become very hot from the preheat flames.

Sharpie markers

These markers are inexpensive and easy to find at any place that sells pens or other office supplies. They work well for layout on clean surfaces, but the marks wear off easily and will disappear when the metal is heated. However, welds on stainless and nickel steels can be adversely affected by the alcohol content in the marker.

Paint markers

Not generally used in layout work, paint markers come in handy for marking metals or other items, such as welding machines and helmets, for identification purposes. Welding over a paint mark will not start a fire, but the organic materials in the paint can contaminate your weld.

Stamps and punches

Center punches are useful for drilling into metal. By center punching the metal first, you give the point of the drill bit an indentation to follow. This prevents the spinning bit from wandering, or "walking," away from where the center of the hole is to be located. For improved accuracy, an optical center punch will magnify your layout lines and allow you to place the center punch indentation in the exact location where the crosshairs line up. Other uses for a center punch include layout and marking spots where parts are to line up. Also, a row of indentations from a center punch is easy to see when following a flat seam in welding. Letter and number sets of stamps are used when more permanent identification markings are required.

CUTTING TOOLS
Drills

Most home shops have a drill or two. Drills are used in woodworking, masonry, and metalworking. Drilling large-diameter holes in thick pieces of steel may require something more than a handheld drill. If it becomes critical to do this work and have the holes be accurate in both location and angle, than either a drill press or magnetic drill should be used. Drill presses are useful for drilling pieces that can be easily moved around or positioned. Be certain to set the drill speed for the type of material (see the section on

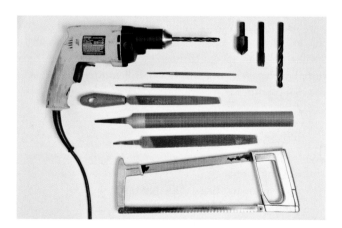

A countersink, tap, and drill are shown at the top right corner. Always use the proper drill size before tapping, and use a tap handle and lots of oil when cutting the threads. Files with handles (like the red-handled one in the center) are safer to use. Files without handles can be dangerous because if your hand slips or the file jams on something, you can be cut or impaled on the sharp, pointed end. *Monte Swann*

21

drilling) and use a drill vise or clamps to secure your work to the table. A drill press chuck spins clockwise, so if the bit should accidentally jam in the piece during drilling, it will spin it in that direction. Keep long objects to the left of your body, so if the work does get caught and spins, it will not hit you. If you need to drill holes in a larger piece of metal that is not easily moved or positioned, then a drill with a magnetic base or Mag drill will be the best option. Mag drills are small drill presses anchored by magnets to anywhere on a piece of steel. Although they are expensive and would only be useful for specific applications, a magnetic drill makes drilling very accurate holes an effortless task.

Drill bits

Drill bits must be made from a harder material than what is to be drilled. Drill bits are made from either plain high-speed steel (HSS) or mixed with cobalt. By adding cobalt, the bits can withstand higher amounts of heat, are tougher, and cost more. There are two cutting edges on a drill bit that are in contact with the material and do all the actual cutting work. These leading edges must be kept in good condition. If these edges are damaged, you can resharpen the bit, if you know how, or replace it. Typical hardware store bits have a 118-degree included angle at the tip, which is fine for cutting through low-carbon steels, aluminum, and copper alloys like brass. For tougher materials like stainless steels, medium-to high-carbon steels, and cast iron, a 134-degree tip is best. Cobalt drill bits already have a 134-degree tip, which works fine for the softer metals, as well.

Drill speed

Setting the rpm of your drill press for the type of material and bit size will make drilling easier and prevent wear and tear on your drill bits. Even if you are using a handheld drill, knowing when to use slow speeds or fast speeds will help. Use a little cutting oil when drilling to help lubricate and protect the drill bit from too much friction. At times it helps to drill a smaller hole first, then increase to the finished size with larger bits. This helps to guide the finish drill and relieves some of the tool pressures.

$$\text{rpm} = \frac{3.8 \times \text{cutting speed}}{\text{diameter of the bit}}$$

Cutting speeds:
Aluminum and brass: 200–300 rpm
Plain carbon steel: 80–100 rpm
Stainless steel and cast iron: 50–70 rpm

If you are using a ¼-inch-diameter drill bit on plain carbon steel, 3.8 × 90 (cutting speed) = 342 divided by ¼-inch diameter (.250 inch) = 1,368 rpm. This is a much lower speed than the 3,800 rpm you would use to cut aluminum with the same size drill bit.

Chop saws (fiber cutoff wheel, carbide wheel)

You will need some method to cut materials to size. A chop saw can do this job well and for little expense. Some chop saws are designed to run at higher speeds. They use fiber wheels, which can be used for cutting carbon steels and stainless steels. Cutting or grinding aluminum with a fiber wheel will clog it up with the base metal. When cutting or grinding again, the wheel will heat up and expand at a different rate than the aluminum imbedded in it. This difference in expansion rates can cause the wheel to fracture and fly apart during operation. Also, particles from the fiber wheel can imbed contaminants into the base material, making it difficult to weld successfully. Other chop saws run at lower speeds designed for carbide-tipped steel blades. Cutting different materials requires changing the blade to one designed for that type of metal. Aluminum can even be cut on a standard power miter saw with a fine-finish 80-tooth blade. Cold saws are similar to chop saws, only they run at slower speeds and use lots of coolant. By reducing friction and flooding the cutting area with coolant, cold saws minimize heat generated when metal is cut. Whenever using a fiber wheel with coolant, shut off the coolant first, allow the wheel to spin off the remaining saturation, and then shut down the wheel. This prevents loading (soaking) the wheel to one side with coolant, preventing balance problems when restarting.

Metal cutting circular saw

Handheld circular saws capable of cutting through metal have recently become available. The saw works much like a circular saw made to cut wood, but with a different blade and a much lower running speed. Separate types of blades can be purchased for cutting steel, stainless steel, and aluminum, and each is rated for a range of thickness.

Chop saws are a good way to cut your metal before welding. This one is equipped with a tungsten-carbide-tipped blade designed for steel. Wear earplugs when using chop saws; cutting with this type of chop saw and blade is very loud. *Monte Swann*

A metal cutting circular saw like this one may be your tool of choice. Be careful to secure small, light pieces, such as angle iron and tubing, with clamps or a vise before cutting. *Monte Swann*

Band saws

The three main types of band saws are vertical, horizontal, and portable. With vertical band saws, the operator must feed the work-piece into the blade. Vertical band saws are commonly used in woodworking and machine shops. Since machine shop

Band saws come in all different sizes. The one pictured here is not the largest, but would be a great size for any shop. The saw is belt-driven and multiple pulleys allow variable speeds, which are useful to cut different types of metals. A coolant pump keeps the teeth of the blade and material cool, allowing for higher feed rates. *Monte Swann*

band saws are for cutting metals, they run at lower speeds than their woodworking counterparts. Cutting metal with a horizontal band saw is easier since the metal is secured with clamps and the blade is fed into the work by the action of the machine. Horizontal band saws come in a wide variety of sizes. Larger saws can cut wider and taller pieces of material. Portable band saws (porta-band) are usually used for handheld, manual cutting of large or odd shaped pieces or frames that cannot be placed in a regular band saw. All band saws give a very clean cut without heating the base metal. A few things to keep in mind when using a horizontal band saw for cutting metal:

- If blade speed is too fast, excessive friction is created. Too much friction causes the blade and work-piece to heat up, causing damage.
- Band saw blade speeds are expressed in surface feet per minute (SFM), which is how far the blade travels past the work in one minute. The SFM is determined by the type of metal you are cutting. Most saws have this information on a wheel or placard on the saw or in the manufacturer's handbook.
- It is safe to run the saw at slower SFM than what is designated for that material. The feed rate will be slower and cut time longer, but it is better to run too slow and not ruin your blade.
- Band saw blades are manufactured with a number of teeth per inch. A 10-tooth blade will have 10 teeth per inch with a spacing of 0.1 inch between teeth. There must always be 2 to 3 teeth in the work-piece at any time. If you were using a blade with 10 teeth per inch to cut tubing with a 0.015 wall thickness, the blade would get caught in the material because not enough teeth were in it at one time. The motor would keep turning, resulting in all the teeth on the blade being stripped off at once, ruining the blade.

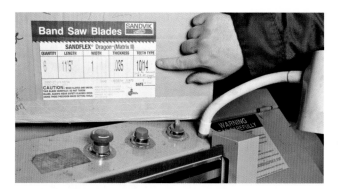

The label on this band saw blade box has a lot of information about the blade, including the blade length, width, and thickness. The number of teeth per inch (10/14) indicates that the number of teeth on this blade changes every inch between 10 and 14. This type of blade is more versatile and will cut through a wider variety of metal thickness. *Monte Swann*

- Wider band saw blades have less deflection when cutting, and have less of a tendency to wander during the cut. Horizontal band saw blade widths are determined by the size of the wheels of the saw. Bigger saws use wider blades. Blade thicknesses are usually 0.035 or 0.025 inch and are for the size of the guides on the saw.
- Use a solidified coolant stick (wax stick) to help reduce friction and heat when cutting on a vertical band saw. Horizontal band saws typically have flood coolant.

Nibblers

Nibblers are handheld electric or pneumatic power scissors with two stationary blades and a third blade that cuts out a strip of metal, severing the piece. Generally they are used for cutting shapes in thinner materials, although heavy-duty nibblers can cut through ¼-inch material.

Tin snips

A good pair of tin snips will also cut through thinner pieces of sheet metal and metal bands. Like using scissors, your hand can become fatigued if you are cutting a large amount of material or thicker pieces of sheet metal.

Reciprocating saw

This power tool may not be the first choice in metalworking, but using a bi-metal blade on a reciprocating saw (Sawzall) will adapt it to cut easily through smaller-sized tubing, pipe, and solid stock.

Hack saw

A hack saw is the manual alternative to a reciprocating saw, costing much less, but requiring more time and effort to cut the same types of material. Use light pressure and cut on the down stroke to get the cut started. Then use elbow grease and the cutting action of the blade to finish.

Files (flat, round, half round)

When you need to remove sharp burrs and corners, or remove that last little bit of material so the parts fit together, a file does the best job. Have a variety of files on hand, including different shapes, such as round, half round, and flat; different teeth, like first cut, second cut, and smooth; and different sizes. Most files are designed to work in one direction only. Look at how the teeth are pointed; applying pressure on the back stroke (when the teeth are pointed in the opposite direction you are pulling) can quickly ruin a file. Small rat tail (round) files are a great way to clean out the nozzle of a wirefeed gun.

GRINDERS

Electric

A 4½-inch angle grinder is the most common type used in metal fabrication shops. Be very careful when using this power tool because spinning discs can come apart and send debris flying at high speeds. I strongly recommend wearing a clear face shield when using an angle grinder with a fiber-based hard wheel or cutoff wheel. If a wheel does come apart, this shield can protect your face from shrapnel, or at least slow it down. After you install a new disc or wheel, point the grinder away from you when first starting it up. This way if something is loose or comes apart, your body will be out of the path of flying debris. Don't breathe in harmful particles and dust thrown into the air by grinding, brushing, or sanding. Be certain to wear a tight-fitting respirator.

- Wire wheels and cup brushes are used for removing rust, paint, and mill scale and for cleaning slag from welds. This is faster and more aggressive than using a wire brush. Stainless-steel wire brushes will not deposit carbon into the base metal.
- Hard wheels are used primarily for grinding down welds. They remove metal quickly and leave a rough surface behind.

The electric angle grinders picture here are 4½ and 7 inches in size. Seven-inch grinders cut, sand, and remove metal more quickly than the 4½-inch, but are heavier, more expensive, and bulky. Six-inch grinders are quickly becoming popular. *Monte Swann*

There are many attachments for a 4½-inch angle grinder. Pictured is a wire cup brush, wire wheel, cutoff wheel, hard wheel, sanding disc, rubber backing pad for sanding disc, and a flapper sanding wheel. *Monte Swann*

Grinding carbon steel produces sparks that can be a hazard in the shop and will cause the metal to heat up from the friction of grinding. Although it is not necessary to wear gloves while grinding, wearing them will protect you from sparks and hot metal. Check the owner's manual for recommendations. *Monte Swann*

- Sanding discs are less abrasive than hard wheels and are used for smoothing surfaces. They come in a variety of grits, and some are specifically designed to sand aluminum. Sanding discs need to have a backing disc for the grinder, which is usually sold separately.
- Flapper wheels are like sanding discs. They also come in a variety of grits, but can last longer than standard sanding discs and don't require the backing disc.
- Cutoff wheels are great for cutting welds or any metal. Although they come in a variety of thicknesses, 0.045 to 1/16 inch is a good size range. Thinner wheels tend to catch less in the groove of the metal being cut and if they do, the wheel tends to shatter. This is actually less dangerous than using a thicker wheel, which is more apt to get caught in the metal and kick back the spinning grinder toward the person operating it.

Out in industry, using grinders to fix mistakes is looked upon as a shortfall in a welder's ability. Skilled welders try to use grinders as little as possible. Although essential for some jobs, using a grinder should not be a substitute for taking the time to do quality work in the first place.

Pneumatic grinders use all the same types of abrasives and wire wheels as electric models, but are less powerful than the electric versions. A straight type (180-degree) die grinder with a carbide bit is especially useful in removing small amounts of metal in tight places, which is usually done manually with a file. A 90-degree angle-type die grinder can be used with 2-inch sanding and Scotch-Brite discs for smaller, more detailed work. Air-powered tools are great, as long as the compressor can supply an adequate amount of pressure for the tool or machine being used, and the air supply stays clean and free of moisture.

METAL FORMING EQUIPMENT

An endless variety of metal forming equipment is available. Depending on the type of work you do, investing in a machine can save time and money. In other situations, you may need only one specialized piece, in which case having the part made at a fabrication job shop may be the best answer. Often shops will have industrial-grade equipment designed for fabricating a specific part you require.

Hydraulic jacks (porta-power) are sold with various attachments that can push and pull metal parts in different ways. This tool is great to have when doing repair work, making alterations, or bringing parts of a project in alignment.

Tube benders are designed to put a radius or arc in structural shapes. Roller-type bending machines use different dies to conform to the shape of the material being formed. Heavy-duty three-roll bending machines are motorized and can handle large structural shapes.

Pyramid rollers come in a wide range of sizes, and have three long steel shafts designed for forming arcs and radii in sheet and plate. Smaller models can be cranked by hand, while larger motorized versions can roll plate several inches thick.

Hand brakes put sharp bends in sheet metal and are rated only to work with metals of certain thicknesses. Thicker materials can be formed using an industrial press brake.

Welding shops benefit from a variety of pneumatic tools and their abrasives. The small cutoff wheel on the left will fit into tight places, carbide and stone bits are center bottom, and small flapper sanding drums are at bottom right. All of these can be used with the die grinders in the bottom left and top right corners. The 90-degree angle grinder at top left is used in a similar way to the electric version, only without as much power. *Monte Swann*

CHAPTER 4
TYPES AND PROPERTIES OF METALS

Is it necessary to be a metallurgist or structural engineer to be a welder? No, but it doesn't hurt. The more you know about metals, their properties and behaviors and the many variations within each group of metals, the more educated your guesses will become when you have to make them, or maybe you won't even have to guess at all.

ELEMENTS AND ALLOYS

At some point in public education, we are introduced to the periodic table of elements in science class. Students fall into two groups after this happens. The first group studies the chart and is fascinated with all its complications. The second group decides they don't want to become scientists. Yet, when we use heat to bring two pieces of metal to their melting point, and then add a filler metal to the puddle, in essence mixing the molecules of all three pieces and watching as they solidify (on a molecular level) into one continuous piece of metal, we are doing science. It's science with torches, arcs, and sparks, but science nonetheless. So we need to know a little in order to understand welding.

The periodic table of the elements lists all the basic substances that make up our material world, each one of which has a unique atomic structure. They are the building blocks, and everything in our material world can be expressed as a single kind atom or a combination of different atoms. For example, pure aluminum (Al) is comprised of only one kind of atom. A molecule, on the other hand, is different because it is comprised of two different kinds of atoms. Water has two hydrogen atoms and one oxygen atom (H_2O). Think of a metal alloy as a water molecule in that it is comprised of two or more elements.

A common alloy is steel. Basic steel consists of iron (Fe) alloyed with carbon (C) and is expressed as the chemical formula FE_3C. Other elements can be added to the mix, but carbon is the most important. In the case of aluminum, pure aluminum is the exception rather than the rule when it comes to products made of that metal. A pop or beer can is made from aluminum alloyed with manganese, even though we just call it an aluminum can. In the chapter on brazing, we will use the filler alloy bronze in the braze welding process. Bronze is a copper alloy, which means it is mostly copper with other elements such as zinc, tin, and iron mixed in.

STRUCTURAL SHAPES

The only way to get a good idea of the variety of materials available today is to visit a steel yard. Spend some time looking around and you'll be astounded at all the different sizes, shapes, and types of materials they have. If you need a certain size for the project you're working on, a steel yard will probably have that shape in a variety of dimensions and can cut it to size for you. Be prepared to pay a cutting charge if you need large quantities or accuracy. The "drop" section of the steel yard is a mixed bag of different pieces left over from the cuts made on new material. There is an advantage to finding what you need in that section—drops are sold at a much lower price.

All steels start life at the mill being hot rolled. Rolled metals, such as sheet and flat bar, have a grain direction along their length. Metal grains become elongated in the direction of rolling and are strongest in that direction. Metals bent or formed parallel to the grain are more likely to crack. Bends made perpendicular to the grain are less likely to crack. Some are left in the hot-rolled condition and are easily identified by the mill scale or heat marks on the surface. Hot-rolled steel (HRS) shapes are dimensionally less accurate, varying in

Metals can be purchased in flat, square, and rectangular bars. Hot-rolled stock has rounded corners and nominal dimensions. Sizes are given in thickness and width dimensions. Lengths are generally around 20 feet. Cold-rolled stock is within a few thousandths of a given size. *Monte Swann*

Angle iron can be purchased in various thicknesses and leg widths. Leg is the name for each flange of the L-shaped piece. Angle iron also comes with two different leg sizes: one wider than the other. *Monte Swann*

For pieces of channel, the middle part is called the web and the flanges attached to it are called the legs. Channels come in standard sizes and a variety of thicknesses. Notice the internal angles on some of the pieces are not 90 degrees. This is important to remember when trying to fit up other pieces to the interior of the channel stock. *Monte Swann*

thickness and width, and are used in basic welding projects. The mill scale does offer a limited amount of protection for rust, although heavy mill scale is considered a contaminant and should be removed from a joint before welding. Hot-rolled sheets are not available in thicknesses less than 0.075 inch.

HPRO stands for hotrolled, pickled, and oiled. A hot-rolled sheet has the mill scale removed by etching and is coated in oil for rust protection. HPRO steel with one "skin pass" is passed through the rollers at a lower temperature once for a smoother surface.

Cold-rolled steel (CRS) sheet and structural shapes are more refined with a better surface finish and tighter dimensional tolerances. Cold-rolled shapes are typically more expensive than hot rolled. There is no mill scale since this group of materials is finished through a cold rolling reduction mill after hot rolling and pickling for scale removal. The grain structure of a material can be greatly influenced by how it is rolled, but the term cold rolled generally does not imply a certain quality, only a finish. There is a higher level of residual stress in cold-rolled steel, which may result in more distortion in parts made out of CRS rather than HRS.

Sheet metal is 3/16 inch or less in thickness and usually comes in 4 × 8-foot sheets. A variety of standard gauge systems are used for sheet metal; the Manufacturer's Standard Gauge for Sheet Metal, formerly known as the US Standard Sheet Metal Gauge, is most common. Other gauges include American Wire Gauge (AWG), Brown and Sharpe, and Birmingham Wire Gauge. To avoid confusion, use decimal measurements to designate thickness. Instead of 18-gauge copper sheet, ask for 0.050-inch-thick copper sheet. Strip stock is also available in sheet-metal thicknesses up to 24 inches wide.

Plate stock is more than 3/16 inch thick. Plate stock comes in enormous sizes, such as 12 × 20 feet, or in flat bars 1/4 inch to 2 inches thick and up to 24 inches wide. When asking for material, designate the thickness, width, and length you require. On structural shapes, length is usually a standard size: either 12 or 20 feet. It may be cheaper to buy a whole 20-foot-long piece and have extra instead of buying only 15 feet. The price per foot is usually higher when purchasing smaller pieces.

A wide variety of metals are sold in sheet and plate form. The pile shown in the top left is comprised of different steel types. Non-ferrous metals (aluminum, brass, and copper) are pictured on the right. The top pieces of steel and aluminum are textured on one side with a diamond plate pattern during the milling process. The piece on the bottom is expanded metal. Steel, aluminum, and brass can be purchased with various patterns of perforation. *Monte Swann*

Tubing comes in square, rectangular, and round shapes. Tubing is designated by the wall thickness and the outside dimensions. Tubing walls range in thickness from .010 inch to more than 1 inch. Tubing is manufactured with welded seams, drawn over mandrel (DOM) or can be made of one solid piece, without seams. Round tubing is measured differently than pipe.

There are three basic types of pipe: standard wall, extra heavy, and double extra heavy. Like tubing, pipes can have welded seams or they can be seamless. Structural shapes with welded seams are usually a lower grade and less expensive than seamless materials. The inner diameter of a pipe is an approximate size until the inner diameter is listed as 12 inches or more. Refer to a pipe schedule chart to find out the wall thickness and inner diameter of a given size piece of pipe. The different schedules refer to the various wall thicknesses available. For example, a 12-inch schedule 40s (standard) pipe has a wall thickness of .375 inch. Two walls equal .750 inch. Therefore the outside diameter is given as 12.750 inch, minus the wall thickness of .750 inch equals an inner diameter of 12 inches, which matches with the given size. However, schedule 80s 12-inch pipe has a wall thickness of .500 inch. Because it has thicker walls, the 12-inch designation is nominal and the final inside diameter is 11.750 inch instead.

NAME AND NUMBERING SYSTEM FOR METALS

Along with the trade names used for alloy steels, there are several different numbering systems for metals. The various specifications include the American Iron and Steel Institute (AISI), Society of Automotive Engineers (SAE), American Society for Testing Materials (ASTM), and the American Society of Mechanical Engineers (ASME).

AISI and SAE have collaborated on their number system: a four-digit index number, which classifies the material by composition or alloy combination. Examples of plain carbon steels are 1010 and 1012. The 10 series has carbon as its main alloy with 0.10 percent and 0.12 percent carbon content respectively. Another example of the AISI-SAE numbering system is 4130.

AISI also has its own classification system for certain alloys, such as stainless steels. The 300, 400, and 500 series are part of this system.

ASTM uses the letter A as a prefix for ferrous metals. A36 (mild steel structural shapes), A242 (alloy steel called Cor-Ten A), and the newer A588 (Cor-Ten B) are steels commonly used in construction. ASTM also uses numbers within an A prefix to designate different grades (classes) of structural shapes, such as bar stock, sheet, plate, rails, and pipe. For example, A 500 is the specification for mild steel cold formed round, square, and rectangular tubing.

Do a little research in books, on the Internet, and at your local steel yards to find out more about these different numbering systems and how they cross-reference each other.

FERROUS METALS

Ferrite is another name for iron. That is why the symbol for iron on the periodic table is Fe. Metals are broken down into two primary groups, ferrous (which contain iron) and non-ferrous (which don't contain any iron).

Steel is made of iron alloyed with carbon. It is the most common metal used on the planet. Steel is also the most recycled material, by volume, more than all other recycled materials combined, including paper, plastic, and aluminum. Steel is recycled so much that the steel in a new car sold today could contain some old steel from a model T, or an old tractor, plow, ship, or demolished skyscraper.

CARBON CONTENT

Making steel is like making a soup. All the ingredients are put into the pot and mixed together, cooked, and served. Today, companies like US Steel use computers and automated equipment to precisely control the amounts of each ingredient added to the steel soup. This has not always been the case.

Before Bessemer's furnace came along, people who made metal soups were known as puddlers. The making of iron alloys and early forms of steel was an apprenticed trade handed down through the generations. As the industrial revolution continued, the need for steel with precisely controlled amounts of carbon was driven by several factors, including the railroad. As westward expansion took place in the United States, the laying of railroad tracks opened the frontier to western settlement. The problem was with the tracks being laid. Before Bessemer, some steel batches had too much of one element and not enough of the other. Tracks were subject to cold and hot temperatures outdoors, and over time would sometimes curve upward away from the railroad bed. When the train came by, these distorted rails, nicknamed cobras, would rip out the bottom of a locomotive and all the cars that hit it at high speed. That is why some passengers would travel with their feet up. Money and time were lost in the process. Along came Andrew Carnegie around 1870 with Bessemer's furnace to solve the problem, establishing the first modern steel mills.

The information given in the following sections is only general information concerning the different types of metals. You may feel overwhelmed, however, it is important to have a basic understanding of materials and how they are numbered and grouped before heading to the steel yard. These number systems are used there and are sometimes written on the metal shapes. Do some research on the Internet and in other text books to help you identify which metals to purchase for your project.

GRADES OF STEEL

Before steel is cast into an ingot at a steel mill, the metal is deoxidized to prevent large gas pockets from forming in the cast metal. There are various degrees of oxidation in steels, depending on how they are processed. Rimmed steels have the least amount of deoxidization. Semi-killed steels have a small amount of deoxidizing agent added to reduce the level of dissolved oxygen. Killed steels are completely deoxidized. The grade of steel can be an important consideration when selecting a filler metal for welding. See the filler rod section in the GTAW chapter for more information.

CARBON STEELS

Small changes in the amount of carbon in steel make a big difference in the characteristics or properties of that steel. Most carbon steel has less than 1 percent carbon.

Steels on the lower end of the medium-carbon range, containing 0.30-0.35 percent carbon, are better suited for welding. Steels containing exactly 0.30 percent carbon are successfully welded with less care and can be semi-hardened. Steels with more than 0.35 percent carbon have a greater tendency to become brittle in the heat-affected zone after welding. Special electrodes (like E7018) and heat control procedures are necessary to prevent cracking.

Low-carbon steel

- Alloy numbers: 1010, 1012
- Carbon content: 0.05–0.15 percent
- Melting point: 2,750–2,786 degrees F
- Uses: many common products, including chains, nails, pipe, screws, and sheets for forming operations (stamped parts)
- Cannot be hardened by heat treatment
- Easily welded

Mild steel (also referred to as low-carbon steel or plain-carbon steel even though it has a slightly higher carbon content than the low-carbon steel listed above)

- Alloy numbers: A36, 1018CRS, 1022 HRS
- Carbon content: 0.15–0.29 percent
- Melting point: 2,700 degrees F
- Tensile strength: 33–40 ksi
- Uses: structural shapes, channel, angle, bar, tubing, sheet, and plate
- Cannot be hardened by heat treatment
- Easily welded

Medium-carbon steel

- Alloy numbers: 1030, 1040, 1045
- Carbon content: 0.30–0.59 percent
- Melting point: 2,600 degrees F
- Uses: axels, connecting rods, shafts, and other lathe-turned pieces
- Can be hardened by heat treatment
- Welded by carefully controlling the cooling rate

High- and very-high-carbon steel

- Carbon content high: 0.60–0.75 percent
- Carbon content very high: 0.76–1.5 percent
- Melting point: 2,462–2,550 degrees F
- Uses for high: dies, car and truck springs, anvils, crankshafts, scraper blades

- Uses for very high: files, woodworking and steel working tools, chisels, metal cutting blades, knives, punches
- Responds well to heat treatment, but large sections cannot be thoroughly hardened by quenching
- Difficult to weld; a great deal of care is required including preheat, special electrodes, and welding techniques; post-weld heat treatment is also required; can be brazed

ALLOY STEELS

Most steel manufacturers will have a family of alloy steels with names like T-1, Cor-Ten, Pitt-Ten, LTV 50XK, and Oregon's A242. Low-alloy steels have alloying elements in small quantities; most common are chromium, manganese, molybdenum, nickel, silicon, tungsten, and vadium. By adding these alloying elements, the physical and mechanical properties of the steel are changed. The higher strength, toughness, ductility, corrosion, and rust resistance make each variety suited to a particular use. The change in properties is remarkable considering the small amounts of alloys used.

Various alloy steels are designed to develop their specific properties during the milling and rolling process, rather than by being heat treated in a separate step. Care must be taken during welding so that these properties are not changed to the extent where the base metal becomes too weak. While the amount of carbon in alloy steels is important, the alloying elements act like carbon, promoting hardenability. A carbon equivalency formula is used to determine the extent of the effects of the alloying elements. No matter which alloys are involved, as hardenability increases, the possibility of cracking also increases.

Low-alloy steel (also known as high-strength steel or high-tensile steel)

- Total alloy content including carbon: 1.5–5 percent (maximum of 9 percent)
- Tensile strength: 90 ksi or higher
- Uses: large structural members, bridges, boats, truck frames, railroad cars, heavy equipment
- Welded with correct electrode (matching the required mechanical properties) and proper welding procedures

4130 (normalized condition) molybdenum steel

- Chromium and molybdenum are the primary allowing elements with 0.30 percent carbon
- Uses: race car and aircraft frames. The high strength-to-weight ratios increase efficiency and performance
- Tubing and sheet up to 0.120 inch is easily welded without subsequent heat treatment
- Filler metal for welding with GTAW process: ER 80S-D2 (first choice) or ER 70S-2
- Joints are required to be cleaned (oxides removed) and degreased within 3 inches of weld area; be careful not to remove too much material, especially on thin wall tubing; materials must be at 70 degrees F before welding.

Tool steels (also known as quenched-and-tempered or tool-and-die steels)

- Alloying elements 10 percent or more, a high-alloy steel
- Alloy designations: D2, H13, A2, A6, CMP 3V, CMP 9V

This group of steels are used to make cutting tools, because they retain hardness at very high temperatures created by the friction in machining operations. They can also be hardened to a greater depth than the same-sized piece of high-carbon steel. There are six major categories of tool steels that have their own classification and letter designation. These steels are difficult to weld because the original mechanical properties of the material tend to get lost.

STAINLESS STEEL

Stainless steels contain alloys of chromium and nickel to increase their resistance to corrosion. Stainless steels will expand 50 percent faster than ordinary steel, so they tend to distort a lot more when heated. There are three types of stainless steels: austenitic, ferritic, and martensitic.

With stainless steel filler rods, electrodes, and base metals, the alloy designation is sometimes followed by an L, which stands for low carbon. The lower amount of carbon reduces the tendency for carbide precipitation to occur. Chrome is normally attracted to carbon, and if given the chance during welding will bond with the carbon instead of the iron, which results in the formation of chromium carbide. Carbon is found in the base metal, filler rods, shielding gas, and grinding wheels. It is critical not only to control the amount of carbon in the material, but also the cooling rate during welding. Slow cooling, especially through the 1,650–800 degree F range, will actually sensitize stainless steel, making it more prone to carbide precipitation. Rapid cooling, on the other hand, will not give the chromium as much of a chance to bond with the carbon.

A stainless-steel weld that turns a straw or copper color is fine. Blue indicates the beginnings of carbide precipitation. Grey or black stainless steel indicates that the chromium precipitated out of the base metal and combined with the carbon, greatly reducing the corrosion resistance of the material.

To weld 300 series stainless, use lower amperages than with mild steel. Make frequent tacks along the joint to hold the metal in place. Fast travel speeds with no hesitation will allow the weld bead to cool as quickly as possible and never preheat the metal. Keep a tight arc length and add filler rod to the molten puddle often to keep the overall heat input as low as possible. Use a purging gas on the backside of joints to prevent contamination from the atmosphere, known as sugaring.

Austenitic (300 and 200 series stainless steel)

- Melts at 200 degrees F less than mild steel

- Alloy numbers: 304, 304L, 308 used in food-grade applications such as restaurant equipment, ovens, and vent hoods; 316, 316L used in structural and marine applications have higher resistance to corrosion
- Easily welded with some care
- These types of stainless contain iron, chromium, nickel, and little or no carbon; for example, 304 stainless contains 18 percent chromium and 8 percent nickel; austenitic stainless steels are not hardened by heat treatment, but can be hardened by cold work (bending, forming metal at room temperature) while still retaining much of their ductility

Ferritic (400 series)

- Alloyed mainly with 11 percent or more chromium; contains little or no carbon
- Considered non-heat treatable and soft
- Uses: building trim, pots and pans; is least expensive

Martensitic 400 and 500 series

- Alloyed mainly with chromium; contains up to 1 percent carbon or more
- Can be hardened by heat treatment
- Uses: knives
- 440C stainless steel is not weldable; 430 magnetic stainless steel comes in a No. 8 mirrored finish

SULFUR STEELS (RESULFURIZED STEEL OR FREE MACHINING STEEL)

Alloy designations: 1112, 1115 free cutting screw stock, 1215, 1244 and 1315, 1330 free cutting manganese.

Sulfur is considered a contaminant in steel and not an alloying element. Sulfur in excess of .055 percent will result in brittle welds. Some steels will have up to 0.33 percent sulfur added to make them easier to drill and machine in a mill. If possible, avoid welding these types of steel. If it is necessary to weld these grades, use an E7016 or E7018 electrode.

WROUGHT IRON (RAW IRON)

Carbon content: less than 0.05 percent.

Uses: ornamental iron work and forge shaping (heating and hammering), old structural shapes.

Wrought iron is a tough, ductile, and fibrous material. It contains a varying quantity of alloying elements and impurities, making it brittle at extreme hot or cold temperatures. Chief among the alloying elements is iron silicate (glass) evenly distributed throughout the material in amounts up to 3 percent by weight.

CAST IRON

Carbon content: 2.5–4.5 percent (roughly ten times as much as most steels).

Grey cast iron is most commonly used because it is the easiest to machine. The term grey describes the surface of the metal when broken, which reveals a dark grey porous surface with tiny flakes of graphite evenly distributed throughout the metal. Other types of cast iron—such as malleable, alloy, and nodular, along with grey cast iron—can be welded. White cast iron has very few applications, is very hard and brittle, and cannot be welded.

Since cast iron is very sensitive to changes in temperature, it should never be heated beyond 1,400 degrees F or it is likely to crack. While the heat from the welding arc will make the molten pool much hotter, it is important that the casting not be held at these welding temperatures for very long. There are two strategies when welding cast iron. First is to uniformly preheat (normally between 500 and 1,200 degrees F) the entire casting and keep it at that temperature until welding is finished. The second is to warm the piece to 100 degrees F and make small 1-inch-long welds. Peen the bead gently in between segments to relieve stresses. Let the bead and base metal cool enough to be able to place your bare hand on it before making another short weld.

In either case, use low amperages to minimize the admixture and residual stresses. Cast-iron pieces need to be buried in dry sand or wrapped in an insulating fire blanket so they can cool very slowly to prevent cracking.

The two common electrodes for cast iron are Ni-55 and Ni-98 with 55 percent and 98 percent nickel content, respectively. The AWS classification for these types of electrodes is Ni-Cl (the suffix Cl designates it as an electrode for cast iron). Repairs on cast iron can more easily be made by braze welding.

CAST STEEL

Instead of being rolled into various structural shapes, large pieces of metal coming out of a steel-making furnace can be cast directly into a specific, irregular shape. Steel castings can be low to high carbon (with no more than 0.50 percent), alloy steel, or stainless steel. The specific type of cast steel can be welded in the same manner as its structural-shape counterpart. For example, a low-carbon cast steel part is easily welded.

FORGED STEEL

Forged steel describes the way a steel part is made. Forged parts include crankshafts, axle shafts, and connecting rods. The metal rod or billet is first heated until it is red hot. Then it is hammered into shape with an industrial forging press. These large power hammers shape the metal and at the same time compress the grains. Forged steel parts are very tough, have a high resistance to fatigue, and are more resistant to cracking than a casting. Forged-steel parts are also made by hand using small forges, anvils, and hammers. Forged-steel parts can be welded with the correct filler metal and correct welding procedures. Repaired parts may not have the same strength as a full replacement part in original condition.

POT METAL

As the name implies, pot metal is made when everything and anything metal is melted down and poured into a mold. Many cast pieces for automobiles, like hardware used in doors, are made using pot metal. Filler alloys are available for welding pot metals, which are added to the base metal when it is heated to a "mushy" state. The chances of success in joining pot metals may be slightly higher using a brazing process instead. There can be no guarantees when repairing pot-metal parts, because of the large variations in base metal compositions.

NON-FERROUS METALS (DO NOT CONTAIN IRON)

Some of the non-ferrous metals not mentioned in this section include gold, silver, and titanium, all of which can be welded.

ALUMINUM

Melting point: 1,217 degrees F.

Aluminum is a recent addition to metalworking and has grown to be the second most commonly used metal after steel. A dirt-like substance called bauxite is mined from the earth and the aluminum is separated by running a huge amount of electrical current through the material. Alcoa is the largest producer of aluminum and a great resource in finding out more about the metal and its alloys.

Aluminum welds are much weaker than the base metal, because the area around the weld will get hot enough to be locally annealed. This means the metal around the weld has become softer, and significantly weaker along the heat affected margins by as much as 30 to 40 percent. For example, 6061-T6 has 45,000 psi tensile strength prior to welding and 27,000 psi tensile after welding (in the as-welded condition).

Copper and aluminum are non-ferrous, meaning they contain no iron. Stainless steel is the only ferrous metal in this photo. Expensive metals, such as stainless steel, aluminum, and copper, can be used to fabricate a great finished project. Because of the expense, be sure to lay out your project carefully and perfect your welding technique and procedures. *Monte Swann*

T6 describes the heat treatment condition of the material. Heat-treated aluminum will have a T followed by a number, while cold-worked aluminum has an H followed by a number indicating its condition.

The only way to improve these properties is to start with a non-heat-treatable aluminum with a zero temper (meaning it has not been hardened by cold work) or perform post-weld heat treatments to heat-treatable aluminums. However, this type of care is only necessary in certain applications. Often the numbers identifying the type of aluminum, like 6061, will be printed right on the structural shape.

1000 series (1100)
- Pure aluminum
- Used in a structural shape or as cladding where good corrosion resistance is required
- Non-heat treatable and low strength
- Can be welded

2000 series (2017, 2024-T4)
- Aluminum alloyed primarily with copper (magnesium or manganese)
- Used in the aerospace industry
- Heat treatable and high strength
- Should not be welded

3000 series (3003)
- Aluminum alloyed with manganese
- Used to make beverage cans and refrigeration tubing; has a shiny appearance and comes in a mirrored finish
- Non-heat treatable and low to medium strength
- Can be welded

4000 series
- Aluminum alloyed with silicon
- Used primarily for a filler metal in welding (4043, 4145) or filler alloy in brazing
- Non-heat treatable

5000 series (5052, 5083, 5456)
- Aluminum alloyed primarily with magnesium (chromium or manganese)
- Used in structural applications in sheet and plate form
- Non-heat treatable and high tensile strength even after softening from welding
- Can be welded

6000 series (6061, 6061-T6, 6053)
- Aluminum alloyed primarily with magnesium and silicon (copper and/or chromium)
- Used in structural applications, extrusions, and furniture
- Heat treatable and high strength
- Can only be welded with the addition of a filler metal; no autogenous welds (weld without filler metal)

7000 series (7075-T6, 7005)
- Aluminum alloyed with zinc and other elements
- Used in the aerospace industry and bicycle frames
- Heat treatable and high strength
- Should not be welded unless the part will receive post-weld heat treatment to restore mechanical properties

Cast Aluminum
Common types of aluminum casting are made of an alloy (212) with 8 percent copper, 1.2 percent silicon, and 1 percent iron. These types of castings can be welded. Other aluminum castings are made from a powdered aluminum that is fused in mold under high pressure. These types of aluminum castings will have a large grainy and brittle appearance on the inside and cannot be welded.

MAGNESIUM

Magnesium is used in die-cast parts and parts manufactured by machining and stamping. For example, some older model lawnmower decks were made from pressed pieces of magnesium. Magnesium possesses many of the same properties of aluminum, such as high thermal conductivity and is welded in a similar way. A unique aspect of magnesium is that it can burn without oxygen. Magnesium fire starters work by shaving off magnesium from a block and using a flint to ignite the shavings. Magnesium can be welded without starting on fire because only small pieces, like chips or shavings, tend to ignite.

COPPER
Melting point: 1,980 degrees F.

Uses: wiring, roofing, tubing for plumbing and electrical components.

Copper was one of the first metals to be used by early civilizations because it exists in its pure form in nature (like gold and silver) and does not need to be refined from an ore. Copper is often alloyed with different metals to produce brass and bronze. It can be welded using the GTAW process and a solid copper wire as a filler rod. Higher amperages will be required because of copper's high thermal conductivity, and the heat of welding will cause the metal to become very soft. Copper and its alloys are more often joined by soldering or brazing.

BRASS
Melting point: 1,652-1,724 degrees F.

Uses: bullet jackets, cartridge cases, musical instruments, ornamental work.

Brass is copper alloyed with zinc, usually between 20 and 40 percent. Certain brasses can also contain tin lead and aluminum for specific purposes. Brass can be joined by welding, brazing, or soldering.

BRONZE

Melting point: 1,566–1,832 degrees F.

Tin, aluminum, and beryllium are alloyed with copper to produce bronze. Zinc may also be present, but is not the principle alloying element in bronze. Bronze/brass metals can be arc welded or used as a filler alloy in arc welding or braze welding.

NICKEL ALLOYS

With alloy numbers 600 and 800, these metals are soft and ductile with a very high resistance to corrosion and oxidation. Nickel alloys such as Monel 400, Inconel, and Hastelloy contain between 60 and 75 percent nickel, 20 percent chromium, and 5 percent iron. These metals are easily welded with the matching filler metal and all the common processes.

METAL IDENTIFICATION

Steel yards will have metals organized and categorized for you. Some yards will use a handheld alloy analyzer to determine the exact content of a mystery metal. This small machine costs as much as a new car, but can reveal if the 2 tons of round stock they just bought is a medium-carbon steel worth 30 cents a pound or an exotic tool steel worth $1.29 a pound.

If the piece of metal you are holding came from a junkyard, how do you know what kind of metal it is? There are several different testing methods for narrowing the possibilities. The function of the part is the first clue; a kitchen knife or tractor rim will be made of a specific type of metal. Next, get familiar with how different metals look. The color, weight, and surface features are all clues in the appearance of the metal, as well as how the interior looks when the piece is fractured. A small magnet will be attracted to most steels, cast iron, wrought iron, and some stainless steels. The harder more brittle metals will be difficult to file or chisel. Metals will throw off different spark patterns when ground on a belt or stone wheel and will react to a flame in various ways. Spend some time doing detective work on mystery metals, do a little research, and you will become an expert in no time.

GRAIN STRUCTURE (MICRO STRUCTURE)

If you ever see photos of a salt grain or snowflake taken under a microscope the geometric patterns are beautiful crystal shapes. Metal is just like salt or snow; it is a crystalline substance, and the grains are the crystals of the metal. Grains can be all different sizes, even up to the point where a piece of metal could be one large crystal. Grains arrange themselves in different patterns, depending upon how the metal is heated and cooled or how the metal is shaped under an applied load.

If a snowflake melts, it turns into water and all the tiny frozen crystals holding it together melt into a liquid. When a metal is melted, the same thing happens. The crystal patterns from when it was solid break down and all the atoms in the liquid mix together. When the metal cools, these crystal grains can reform in different ways, depending upon the rate of cooling.

When water freezes in an ice cube tray, you may notice that it freezes first at the edges of the tray, and last in the center. Lakes freeze over in the same way, with ice forming at the shore first. The same concept is true when molten metal freezes (solidifies). The coolest spot in the weld is where the still-solid base metal meets the molten metal. This is where the crystals will form first. The center of the molten puddle will solidify last.

There are many different grain structures; some are more favorable than others. In the case of alloys like steel, each of the elements has a different melting point and solidifies at a different temperature. The carbon atoms in steel are free to move around when melted, but can be distributed in different patterns among the iron atoms when the whole thing solidifies. These different grain structures are known as ites: austenite, ferrite, pearlite, cementite, and martensite.

PROPERTIES OF METALS

The usefulness of a particular part depends upon if that part can withstand the forces acting upon it. For example, metal that is stamped into a car door panel or drawn into a cooking pot needs to be softer than the metal used to make a drill bit or more pliable than an armor plate that must withstand the impact of a bullet or explosion. Climate and environment can also be a factor. Steel used to build an oil rig in the Gulf of Mexico will posses different properties than metals used for pistons and engine blocks or a pipeline in northwest Canada. All the properties of the base and filler metals are carefully considered before any of these high-cost projects are undertaken.

The heat from welding affects metals in a variety of ways, and often changes their properties. Metallurgy is the study of metal properties. Professionals make a career studying this science, with a long history dating back to the alchemists. Like we discussed earlier, you don't have to be a metallurgist or structural engineer to be a good welder, but it is important to have a basic understanding of these concepts.

- *Chemical properties* describe how a metal reacts with other elements in the environment. For example, the oxygen in the air and water (H_2O) combine with the iron in steel to form iron oxides, more commonly known as rust. The act of these two elements chemically combining is called oxidation. Corrosion describes the deterioration of a material. Salt corrodes aluminum, eating away at the metal without a new material being formed.

- *Physical properties* are unique to each kind of metal. For example metals liquefy at different temperatures; it takes a lot less heat to melt lead than to melt iron. The melting point for each type of metal is different and can be used to our advantage. Tungsten has one of the highest melting points, so it is used as an electrode in the GTAW welding process because it can withstand the heat of the welding arc. Before welding can take place, both the base metal and filler metal need to be very near or above their melting points. This is a critical factor in successful welding. Cold welds will lack proper fusion between metals. Like water, metals also have a boiling point. In welding, reaching the boiling point of a liquefied metal is never good.

Metals also conduct heat and electricity at different rates. Copper is used for electrical wiring because it is highly conductive and economical. Silver and gold are better conductors, but their use is limited due to rarity and cost. Most metals are electrically conductive—electrons can easily pass through them—allowing us to use an electrical current to weld them.

Thermal conductivity is measured by the rate at which heat passes through a material. This type of conductivity can affect the rate metal heats up, how much heat can be concentrated in one area, and the rate of cooling. Copper has high thermal conductivity. High amounts of heat are required to create a molten puddle, since most of the heat from a torch or welding arc is moving quickly away from the spot being heated and into the rest of the work piece. Physical properties are specific to the type of material and are not related to an applied force.

- **Mechanical properties** are exclusively related to an applied force. Mechanical properties are those that determine the behavior of metals under an applied load. Heat can easily change the mechanical properties of certain metals that are heat sensitive. Mechanical properties are studied under a range of temperatures from below freezing to the melting point of a metal.

Load is an external mechanical force applied to a part, welded or otherwise.

Static loads remain constant, like the weight of a building bearing down on the beams and girders. Impact loads are applied suddenly or intermittently. Think back to that bullet impacting a steel plate or the spinning teeth of a rock crusher. Cyclical or variable loads vary in time and rate, but without the sudden change that occurs with an impact. A highway bridge is under a variable load. Not only do the beams and trusses need to hold the weight of the bridge, but also the traffic driving across it. Semi trucks weigh much more than cars, and the number of vehicles is much different during rush hour than at 3:00 a.m.

Loads can be applied in different directions, as well. The five types of mechanical loads that can be applied to a part are tension, compression, shear, torsion, and flexing. In most cases applied loads will be combinations of these five types.

Stress is the internal resistance of a material to an externally applied load. When someone is stressed out, they feel pulled in all directions. When metal parts are stressed, they are being pulled, crushed, twisted, and bent. A high enough level of stress on a part will cause it to bend out of shape or break down.

Strength is the ability of a metal to withstand some kind of force or load without breaking down. Metals with high fatigue strength can withstand cyclical loads.

Tensile strength is one of the most important factors in determining which metals are used for structural members, machine parts, and pressure vessels. Tensile tests are commonly performed on base metals, weld metals, and welded joints to indicate their strength. To do this test, parts are pulled in opposite directions. During testing, a good weld will be stronger than the plate and break at the base metal. A poor weld will come apart either at the weld or right next to it in the heat-affected zone.

Tensile strength is a common measurement for steels and welding rods (electrodes). Metals and welds with high tensile strengths can withstand higher loads and a larger amount of tensile stress before breaking down. Right before becoming completely stressed out and breaking down, metals will usually change shape and, in the case of tensile loads, stretch lengthwise.

Toughness is the ability of a metal to absorb energy without breaking when a load is applied slowly. Metals with a low toughness are brittle, while metals with high toughness are ductile. Notch toughness is the same measure of a metal's ability to absorb energy, only in a material with surface flaws, such as welds with excessive convexity or undercut. The harder and colder a metal is, the less notch toughness it will have.

Strain refers to a metal changing its dimensions when a load induces stress. *Elastic strain* (deformation) takes place when the material changes shape, but is able to return to its original dimensions after the stresses from a load have been removed. Some metals are like a rubber band—they can withstand a lot of stress and still return to their original shape, like a car spring or sword. *Plastic strain*

This is a tensile pull specimen. This tensile test was successful because the fracture happened in the base metal and outside the weld and heat-effected zone. Notice the elongation and narrowing of the metal next to the fracture. The dented pattern in each end of the specimen is caused by the pressure of the clamps. *Monte Swann*

Over-welding distorted the base metal of this part. Excessive amounts of weld metal deposited in the tee joint caused a large amount of shrinkage stresses. Small fillet welds on both sides of the joint would help alleviate this problem. *Monte Swann*

are names of tests performed to determine a metal's hardness. Hardness is a desired property in tooling used in machining and saw blades. Hardening steel will increase its strength and resistance to scratching, cutting, and abrasion but lower its ductility. Hard-facing welding electrodes are used to coat the surface of parts that see a lot of wear, such as the bucket of a front-end loader or the blade of a road grader.

EFFECTS OF HEATING METALS

Welding is all about controlling the temperature of the base metal. The amount of heat in welding will depend on a lot of factors, but controlling the heat is the key to success. All the mechanical properties mentioned are directly influenced by the way in which the metal was heated and cooled before, during, and after a weld. The rate of cooling is a critical element in determining if the weld is ductile or brittle.

Heat-affected zone (HAZ)

Even though the molten metal in a weld pool is at a very high temperature, the base metal even 1 inch away from the weld is substantially cooler. The heat-affected zone (HAZ) is the metal directly adjacent to the weld bead, which is heated very close to the base metals melting point. The HAZ on a ½-inch-thick plate is approximately ⅛ inch wide with a grain size ranging from fine to coarse. The cooling rates in the HAZ are highest because the surrounding base metal rapidly absorbs the heat from the weld, creating an effect known as contact quenching. The HAZ can change in size and the crystalline micro structure of the grains can be altered depending on the total heat from welding, metal thickness, rate of cooling, and alloying elements in the metals, like the carbon content in steel.

refers to when a metal is permanently deformed by load-induced stress.

In a press brake, metals are placed between two hard steel dies and bent at a certain angle. An example of this process is the sheet metal on a refrigerator door, which has been formed in a machine that bent the metal into a new shape. The metal has been plastically deformed to make a useful part. Metals such as stainless steel are ductile enough to withstand the strain of being bent without cracking or breaking. But, put a hard piece of tool steel under a high enough load and it will shatter like glass.

Ductility is the measure of the metal's ability to yield plastically under load, rather than fracturing. Ductility is a good thing in welding, because ductile welds are better able to withstand loads that stretch, compress, twist, and/or bend the metal. *Yield strength* is the measure of the stress within a metal that is enough to cause plastic flow and make it bend. Ductility is important in metals that are used in forming operations. Punch presses and press brakes are used to manufacture formed parts, like automobile body components and refrigerator doors.

The amount of ductility is dependent on the type of metal, its grain structure, and its temperature. Metals are more ductile at high temperatures and less ductile at low temperatures. Metals ductile at room temperature may fail in sub-zero conditions.

Brittleness is the opposite of ductility and is also called low ductility. Like the tool steel mentioned above, a brittle material will show practically no signs of bending or permanent distortion before fracturing. A ductile base metal with a brittle weld is a very dangerous combination when put under any type of load.

Hardness, also called stiffness or temper, is the resistance of a metal to plastic deformation, usually by indentation but also by scratching, abrasion, or cutting. Rockwell and Brinell

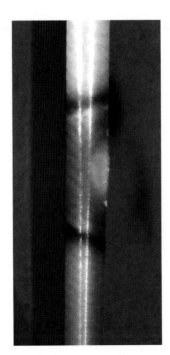

This square metal bar was heated with an oxy/acetylene torch until one side began to glow orange at about 1,800 degrees F. Notice the temper colors indicating the difference in temperatures between the hottest and coldest parts of the bar. When this bar begins to cool, the shrinkage stresses between the heated and non-heated areas will cause it to distort and bend toward the heated side. *Monte Swann*

While it is true that the weld metal can be stronger than the base metal, if the mechanical properties of the HAZ are poor, the welded metal will come apart.

Rapid cooling (quench effect)

Certain crystalline grain structures form when metals are cooled rapidly. The rapid cooling of metals—especially medium- and high-carbon steels, tool steels, and cast iron—will make the weld metal harder and more brittle. This will increase the chances for cracking, especially when a load is placed on the part.

Submerging a hot weld in water or oil is called quenching and is the fastest way to rapidly reduce the temperature of a metal. Large pieces of thick metal will absorb heat rapidly from the weld and HAZ, causing contact quenching. Even metals that cannot be hardened by heat treatment, like mild steel, will crack if welded in cold weather. Always heat up the base metal to 50–70 degrees F before welding.

Slow cooling

When you think of steel beginning to solidify, it's like the end of the day at a junior high school. The students are atoms of iron and carbon (in liquid form), excited and bouncing around, going here and there. The busses waiting for them are the crystals forming in the solidifying metal. Your goal is to get the right kids on the right buses. If this is done quickly, they will all cram together with their friends. You'll end up with too many kids on some buses, while others will be empty. If it is done in a slow and organized way, the right amount of kids will get on each bus. With steel, the right amount of carbon, iron, and other elements will be evenly distributed among the grains. Slow cooling promotes strength and ductility because the grain structure that forms has an evenly distributed amount of carbon.

The slow cooling of the weld and HAZ reduces the tendency for a metal to become brittle and crack under load. The use of preheat will reduce the cooling rate of the weld and surrounding metal. After welding, a part can be buried in kitty litter or dry sand, or wrapped in a fire blanket to cool more slowly than if it were left out in the air. Some metals are highly sensitive to the cooling rate and need to be heat treated after welding.

ADMIXTURE

When fusion takes place between the base and filler metal, some of the base metal is mixed into the molten weld bead. This combination of metals is called the admixture and often has different properties than either the base or filler metal alone. The admixture of a weld can vary depending on the amount of heat, depth of fusion, travel speed, and type of filler metal used.

HEAT TREATMENTS

There is no one magical heat treatment. In fact, there are so many varieties and methods to heat-treating materials that it is a science. Early metal workers like blacksmiths constantly heat treated the parts they would place in the forge, hammer and shape, quench, and place back in the forge for tempering. Today, metals like steel and aluminum are put into categories of heat treatable and non-heat treatable depending on the different alloys and the quantities of those alloys. Many commonly used metals will not require heat treatment after welding. Other metals, such as quenched and tempered steels or car springs, are already in a heat-treated condition. Structural shapes made from metals that are hardenable by heat treatment or cold working (bending and rolling) will have letters and/or numbers indicating the condition of the metal. Welding on these metals creates localized heating and radically alters the metal's mechanical properties. Care must be taken when welding metals that can be hardened by heat treatment. After these metals have been welded, the condition of the metal is described by as welded, indicating no subsequent heat treatment has taken place.

Often heat treatment is a tradeoff. It can increase the strength and hardness of a metal and at the same time make it more brittle and less ductile. You can perform a basic heat treatment on most hand tools and cutting blades. It will take some trial and error and close observation to work. The following procedure illustrates the steps to follow, using a screwdriver as the test mule:

1. Use an oxy/acetylene torch with a No. 0 tip to heat the end of a slot-head screwdriver. Move the torch back and forth, heating 1 inch of the blade until it turns cherry red. An older screwdriver with a large shank will work well for this exercise.
2. Immediately plunge the blade into a bucket of room temperature water. Swirl it around so the quench will be uniform.
3. Sand, buff, or wire brush the discoloration and scale from the blade until it is shiny. Note: If you try to use the screwdriver at this point, it will shatter because it has been fully hardened.

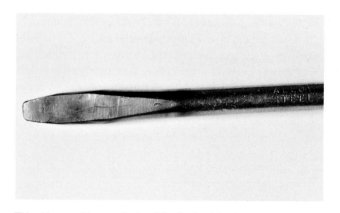

This old screwdriver worked well for the heat-treatment exercise. Notice the shank has alloy steel embossed in it and the blade has cracked in the middle from rapid quenching. *Monte Swann*

4. Using a less intense flame, pass it back and forth across the shank, 1 inch up from the tip. You will begin to see the temper colors form very quickly. When this occurs, gently move the flame toward the end of the blade, using the torch like a paintbrush to pull or draw the colors up to the tip.

5. The colors will move very quickly toward the thin section of the blade. You may need to remove the heat and watch the colors creep to the tip.

6. As soon as the end turns violet or purple, quench it again immediately.

7. You might apply too much, or too little, heat. It may end up being the wrong temper color or turn grey from overheating, which is too soft for cutting tools. If you overheat, you will need to reharden and retemper the steel, being more conservative with the torch. If it is left too hard, retemper to a higher temperature.

PREHEATING

The main function of preheating a base metal is to slow the cooling rate. Usually, the heat of welding is rapidly transferred out of the base metal and HAZ, especially on thick sections of metal. If the base metal has already been heated (usually between 300 and 400 degrees F) it takes more time for the weld metal and HAZ to cool, allowing time for a more favorable grain structure to develop and lowering the amount of internal stresses.

ANNEALING

Annealing is a procedure used to soften heat treatable metals. Annealing refines the grains and removes stresses in the metal. Steel is heated 100 degrees above the temperature at which the grain structure changes. For steels with 0.83 percent carbon content, 1,333 degrees F is the transformation temperature. The critical temperature, or transformation temperature, is different depending on the metal. Steels held at this temperature will uniformly redistribute carbon among the grains. In order for annealing to take place, metals are held at this high temperature for a certain period of time and slowly cooled in an oven or packed in lime ashes. Annealing is a process done before quenching and tempering. Welded parts are usually not annealed since they would distort from the high temperatures involved.

QUENCH HARDENING

Before quench hardening can take place, metals are brought up to an annealing temperature. They are heated just long enough for the grain structure to change (becoming an austenite structure), but not too long for the grains to grow large. The hot metal is submerged in water or oil and swirled in a figure-eight pattern. At this point, the metal has been fully hardened all the way through. After quenching, the metal is very brittle and easily fractured. Quench hardening is different than case hardening, in which only the surface layer has been hardened.

TEMPERING

Tempering is similar to stress relieving. A hardened metal is reheated to a temperature below the transformation or critical range and air cooled or requenched. This process reduces the metal's hardness/brittleness and increases its toughness along with relieving some of the internal stresses. Tempering provides a good balance between the mechanical properties of a high-strength part. The higher the tempering temperature, the lower the hardness and tensile strength. Normal tempering temperatures are between 425 and 620 degrees F.

STRESS RELIEVING

Non-uniform heating from a weld results in non-uniform expansion and contraction causing distortion and internal stresses to develop. Stress relieving is performed on medium- to high-carbon steels, alloy steels, and heat treatable metals. The entire piece of welded metal is heated uniformly, hot enough to relieve internal stresses but not too hot to change the grain structure. The piece is then cooled at a uniform rate.

MAKING REPAIRS

The main goal in welding or brazing two pieces of metal together is to have them stay together. Metallurgists and structural engineers have spent more than a century closely studying how to accomplish this goal in new and better ways. Some of the benefits of their research have been handed down to us in the form of new and better metal alloys and technologies to join them. The Internet provides easy access to a vast amount of information on the subject. Even with all of this knowledge, it can be challenging when someone hands you a part they want put back together. I am constantly asked to repair metal items that break. When I am asked to fix things, there are a few questions I ask first.

1. What type of metal is the part made of?
   In some cases, the part is made from a material that cannot be welded. In other cases, brazing would be a better option because of the material's sensitivity to high temperatures. Often, the person asking you to do the job will have no idea. In that situation you may have to do some detective work before you begin. If you take an uneducated guess and go right to it, you may get lucky and have the weld hold. But this is not a good way of making something usable again. Read about different types of metals and their applications; examine different materials at a steel yard or junk yard. If necessary, make educated guesses.

2. Why did it break and can I prevent it from breaking again?
   Before I fix something I try to determine why it broke. There are many reasons why welds break, but the common cause is overloading. Load is defined as an external mechanical force applied to an object. For example, a car's axle must support the entire load of the

Heating Temperatures and Colors of Steel

Color	Temperature (F)	Process
White hot	2,800 2,700 2,600 2,500 2,400	Welding range High-speed steel hardening (2,250–2,400 degrees)
Yellow white Bright yellow	2,300 2,200 2,100	
Yellow / lemon	2,000	
Lemon / orange	1,900	Alloy steel hardening (1,450–1,950 degrees)
Orange	1,850	
Orange / red	1,800 1,700	
Light cherry red Bright red	1,600 1,500	Carbon steel hardening (1,350–1,550 degrees)
Cherry red	1,400	
Dark red	1,300 1,200 1,100	Critical or transformation temperature for 0.083 percent carbon steel (1,333 degrees)
Very dark red Blood red	1,100 1,000	
Faint red	950	

Temper Colors for Carbon Steels

Color	Temperature (F)	Suggested uses for carbon tool steels
Light straw Pale yellow	425	Steel cutting tools, files, paper cutters
Dark straw Deep yellow	462	Punches and dies
Gold Yellow brown	490	Shear blades, hammer faces, center punches, cold chisels
Purple/bronze	500	Axes, wood-cutting tools, striking faces of tools
Violet/purple	540	Springs and screwdrivers
Pale blue	580	Springs
Dark blue	590	
Steel grey	620	Cannot be used for cutting tools

vehicle, distributing it to the wheels. Engine mounts on a car must be designed in a way to support the load of the engine, both its weight and vibration. In a crash, the structural parts of a car's panels and frame will buckle and deform. The impact of the crash is the load being placed on the material. The cause of a broken weld or metal part is obvious when accidents happen. But if the weld breaks under normal service conditions, it may be a problem with matching the base metals and filler metals properly, or a problem in not following proper weld procedures for certain types of materials. Some of the factors to consider when making a repair are as follows:

- Parts break because the last person did a poor job of welding in the first place. If you can see that the weld was misplaced on the joint and did not hold the part together, it's an easy fix to reweld it correctly. Missing welds may or may not be as obvious to spot.
- Keep in mind as a general rule: more weld is not necessarily better and may in fact be worse. Make careful decisions. Don't jump to the conclusion that adding more weld metal will make a joint stronger.
- Welds break because of corrosion after years of service. Welding on rusty parts is never a good idea. Rust and corrosion can thin the materials down to a point where they will not carry the load required to perform their function. Rusted or corroded areas should be removed down to sound metal before welding, if there is any left to begin with.
- A part's design, or the material thickness used to make the part, was inadequate for the loads

being placed on it. We live in a time when the machines, household items, and tools/parts we buy are engineered and manufactured with the least amount of material required. Often I will add some reinforcing members to an under-engineered part to prevent the problem from happening again.

- If the part has multiple welded joints and one of them breaks, after fixing it, look for the next weakest link. It is easier to do some preventative repair and reinforcement while you have tools out and the machine set up and running, instead of having to start over again.

3. Is it worth fixing?

There are two considerations here: economic value and liability. Many times people want things fixed because it is cheaper for them to have you do the work. Once you have some repair experience, you'll know how much time and effort can be involved in a project. Ask: is it a one-of-a-kind, hard-to-find item, or can a replacement be purchased at a reasonable price? If it is a quick fix on a common item, it may be worth doing to save that person some money. If they are your friend or someone who you might need a favor from, then it may well be worth fixing. The bartering system can be a great thing, but save yourself some potential embarrassment by not taking on projects above your skill level.

The second consideration is liability. For example, I refuse to weld on the forks of a fork lift. These critical parts carry all the weight and can be purchased new at a reasonable price. If repaired forks fail, it can have dire consequences for anyone in the way of falling materials.

CHAPTER 5
JOINT DESIGNS, WELDING POSITIONS, AND DISCONTINUITIES

Put two pieces of metal together. Sounds simple, and to most people it is. Give ten different people the same two pieces of metal and they probably will put them together two or three, or maybe even ten different ways.

There are two main categories of welds: fillet welds and groove welds. Fillets are most often used to join metals meeting at 90-degree (or similar) angles, while groove welds are most typically used to connect butt joints. A groove weld deposits welding material into the groove of a joint, while a fillet weld adds material primarily to the surface of a joint.

Bear in mind there are different types of joints (see section on joints below). A tee joint can be bonded with a fillet and/or a groove weld. A butt joint is connected with only a groove weld—a butt joint cannot have a fillet weld applied to it.

Your performance as a welder can improve just by knowing a little about the different types of joint designs and the types of welds applied to them. It is difficult, if not impossible, for even the best welder to get an adequate weld on a poorly designed joint.

FILLET WELDS

A fillet weld sits on top of perpendicular surfaces and is not designed to penetrate into a joint. Good fusion must take place between the weld and base metal, especially at the root of the joint. But think of a fillet weld as something sitting next to the two 90-degree surfaces; it is essentially a right triangle in section. Fillet welds are the most common type of weld. A good example is one tube fitting into the other in a copper water pipe or car exhaust. The two pieces overlap, and where the edge of the larger-diameter tube meets the outside wall of the smaller-diameter tube, that is a lap joint where a fillet weld can be applied.

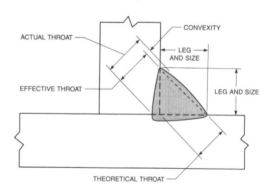

(A) CONVEX FILLET WELD

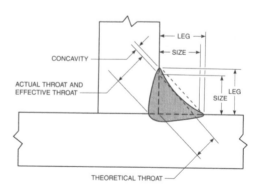

(B) CONCAVE FILLET WELD

AWS A3.0:2001, Figure 25, Reproduced with permission from the American Welding Society (AWS), Miami, FL USA

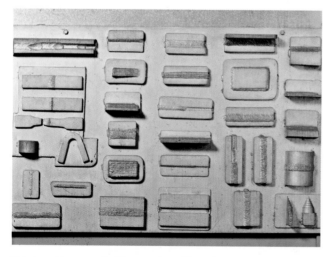

This board is covered with example welds and test specimens. The more welds you look at, the better able you are to evaluate your own welds. *Monte Swann*

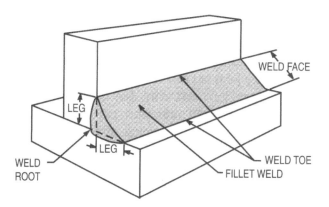

Welding Inspection Technology, Fourth Edition, Figure 4.22, Reproduced with permission from the American Welding Society (AWS), Miami, FL USA

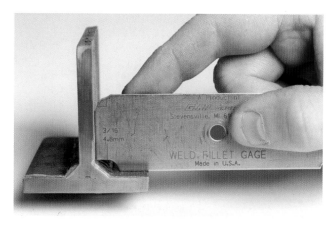

A set of fillet-weld gauges are used to measure fillet-weld sizes. Since the fillet weld has a slightly convex contour, this side of the gauge shows the leg size is 3/16 inch. If the fillet weld is concave, the top side of the gauge is used to determine the size of the weld. *Monte Swann*

Face

The face of a fillet weld can have different contours. Depending on how it is welded, the contour can be flush, convex, or concave. The face can also describe how the weld was made by how the ripples appear on it.

Legs

Typically the legs on a fillet weld are the same size. When they are the same size, the throat of the weld is a sufficient size, which means the weld is stronger. When the legs are not the same length, the fillet weld has unequal legs. The smaller size of one of the legs can greatly reduce the size of the throat and reduce the strength of the weld. In some cases, the joint will have unequal legs, like when joining two metals of different thicknesses. For example, welding sheet metal to a piece of 3/8-inch-thick plate will require a fillet weld with unequal legs.

Toes

All fillet welds have two toes. (The toe is the point where the base metal meets the filler metal.) Look at the toes of your weld for good fusion and a smooth transition. Undercut and overlap problems occur at weld toes.

Root

The root of a weld can be found at the root of a joint. The joint root is where two pieces of metal are closest to each other. For fillet welds, it is important to have complete fusion at the root of the joint. The weld metal should have completely fused both pieces at the point where they come closest together in order to have maximum strength. Proper fusion at the root of a fillet weld is almost impossible to detect visually.

Throat

Engineers calculate the strength of a weld by measuring the throat size. There are different ways the throat of a fillet weld is measured: actual, effective, and theoretical. The difference between throat sizes is not as important as knowing that a fillet

weld is strongest when it has the proper throat dimension. If you measure from the root and make the shortest line from there to the face, that is the throat. Unequal legs and concave weld profiles will reduce the amount of throat in a fillet weld and reduce its strength. In fact, if you need to cut apart a fillet weld, cut its throat. Cutting through the throat of a fillet weld is the fastest way to get two pieces apart because you are cutting out its strength.

FILLET WELD SIZE

Fillet weld gauges are used to measure fillet weld sizes. The size of a fillet weld with a flush or convex contour is equal to the leg size. This is not the same for fillet welds with concave contours. The size of a fillet weld need only be as large as the thickness of the base metal.

A common mistake beginning welders make is to overweld their projects. Avoid making welds larger than necessary. Larger welds or additional welding do not necessarily make the part stronger, and in some cases can make it weaker.

Under-sized fillet welds can also be a problem because things come apart if not enough weld has been put down on the joint. People who have had their welds fail tend to add more weld in an attempt to fix the problem so the parts will hold. Hobby or farm welding is not professional welding and is usually not pretty. Avoid this style of welding and learn the correct way. It may surprise you how little weld is necessary for parts to hold if the welding is done correctly in the first place.

For sheet metals 3/16 inch or less in thickness, make your fillet weld the same size as the base-metal thickness. On some thinner sections, when using a semiautomatic process such as GMAW with an 0.035-diameter wire, it may be necessary to make the fillet weld size slightly larger than the base-metal thickness in order to achieve good fusion between the weld and base metals. If you are welding two different metal

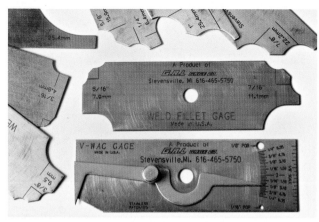

Shown at the top is a set of fillet weld gauges used to measure welds with both concave and convex contours. Below it is a V-Wac gauge used for measuring undercut, underfill, and the amount of reinforcement on groove welds. It can also be used to measure undercut on fillet welds. *Monte Swann*

thicknesses, the size should be no larger than the thinner of the two pieces.

For plates:

¼ to ½-inch thick use a ³⁄₁₆-inch minimum size fillet weld

⁹⁄₁₆ to ¾-inch thick use a ¼-inch minimum size fillet weld

⅞ to 1¼-inch thick use a ⁵⁄₁₆-inch minimum size fillet weld

GROOVE WELDS

Groove welds are the only kind that penetrate into the joint. Butt joints, where the edges of two pieces of metal butt against each other, have only groove welds applied to them. Groove welds can be made on sheet metal, on plate, and when connecting the two ends of a piece of pipe or tubing. Groove welds can also be applied to the other types of joints, especially if CJP is required.

Groove weld size

The size of a groove weld is determined by the thickness of the base metal. A full-size groove weld on ⅛-inch sheet metal is ⅛ inch. Groove welds can be smaller than the thickness of metal, but never larger. A groove weld may be applied on a tee joint for full penetration. Afterward, a fillet weld may be applied either to the same side as the groove weld and/or the other side of the same joint to increase its strength.

Depth of bevel

Metals ⅛-inch thick or larger are typically beveled before welding. Preparing the joint by beveling increases the depth of penetration of the weld. Beveling can be done with a grinder on metals ¼-inch thick and less. Pieces ⅜ inch and over will need to be beveled with a torch, cut in a horizontal

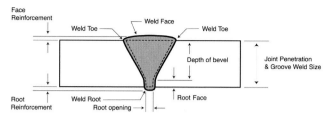

Groove-weld profile diagram. *Rob Lindgren*

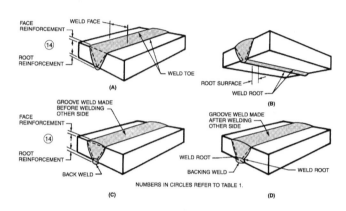

AWS B1.10:1999, Figure 9, Reproduced with permission from the American Welding Society (AWS), Miami, FL USA

band saw, or milled. The more penetration a groove weld has into a joint, the stronger it will be. The depth of bevel will depend on if you need a full-penetration weld, if the joint is to be beveled on both sides (double bevel or double V), or if you need extra metal at the root of the weld to carry the heat required for full penetration.

The bevel angle is an important consideration. Two beveled plates put together form a V groove and have an included angle (groove angle) of the two beveled angles added together. A 25- to 30-degree bevel angle is a typical joint preparation for a V groove, which would have a 50- to 60-degree included angle (groove angle).

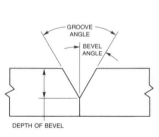

AWS A3.0:2001, Figure 6c, Reproduced with permission from the American Welding Society (AWS), Miami, FL USA

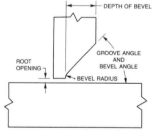

AWS A3.0:2001, Figure 6e, Reproduced with permission from the American Welding Society (AWS), Miami, FL USA

Root opening

If a gap is left between the pieces of metal, this is called the root opening. Root openings are typically used in butt joints to achieve adequate penetration, but can be used in tee joints for the same purpose. On metals thinner than ⅛ inch, the proper root opening will make the joint easier to weld. However, on thinner pieces ⅟₃₂ inch or less in thickness, you may not need any root opening to achieve full penetration. Typically, the smaller the gap, the less deep the penetration in the joint. The larger the gap, the more difficult it can be to weld, sometimes requiring one or both edges to be built up with surfacing welds before the final welding can be done.

Depth of penetration and CJP

Some welds need CJP to withstand the amount of loading and forces applied to them. Other welds will hold up just fine even if they only partially penetrate the joint (partial penetration groove welds). For example, welding a tubular frame for a barbeque cart will only require a fillet weld all the way around the tubing at each joint, especially any fish-mouth joints that come together at various angles. If the tubing is butted up at the ends, then a groove weld is necessary, and so depth of penetration becomes an important consideration.

Face reinforcement

Although a standard term, face reinforcement provides no additional strength to most groove welds. Some face

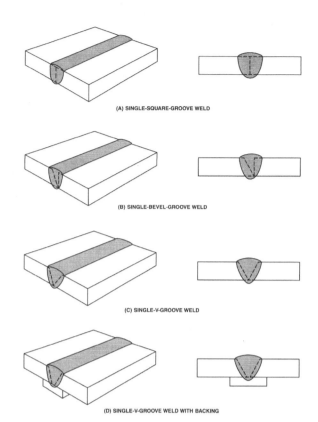

(A) SINGLE-SQUARE-GROOVE WELD

(B) SINGLE-BEVEL-GROOVE WELD

(C) SINGLE-V-GROOVE WELD

(D) SINGLE-V-GROOVE WELD WITH BACKING

AWS A3.0:2001, Figure 8, Reproduced with permission from the American Welding Society (AWS), Miami, FL USA

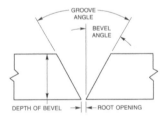

AWS A3.0:2001, Figure 6a, Reproduced with permission from the American Welding Society (AWS), Miami, FL USA

This is a single V-groove weld on a butt joint in the horizontal position. When the depth of bevel is equal to the material thickness it is sometimes called a feathered edge. *Monte Swann*

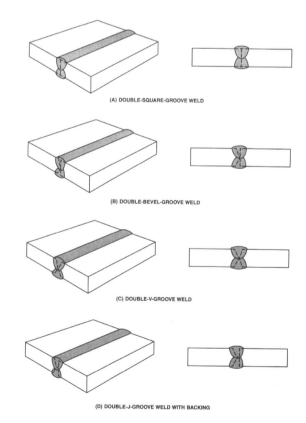

(A) DOUBLE-SQUARE-GROOVE WELD

(B) DOUBLE-BEVEL-GROOVE WELD

(C) DOUBLE-V-GROOVE WELD

(D) DOUBLE-J-GROOVE WELD WITH BACKING

AWS A3.0:2001, Figure 9, Reproduced with permission from the American Welding Society (AWS), Miami, FL USA

43

reinforcement will help a groove weld all the way around on tubing if it is not a full penetration weld. Other than that, face reinforcement will not provide additional strength to a joint. In fact, excessive reinforcement can be detrimental to the strength of the joint. A highly convex face on a groove weld—more than ⅛ inch in height—creates a notch effect at the toes of the weld.

Underfill

If, for example, a V-groove joint is supposed to have CJP, and the weld face has a concave appearance, the joint is underfilled. Only a groove weld can be underfilled. A similar problem in a fillet weld would be called insufficient throat. Underfill is different than undercut. Undercut is in the base metal and is caused by improper welding technique. Undercut is not caused by a lack of deposited filler metal. See the section on weld discontinuities for more information on undercut.

PLUG AND SLOT WELDS

If you have two overlapping pieces of metal, and you don't want to see the weld, or need clearance for another piece where a regular weld bead would be in the way, a plug weld is your best solution. Plug welds are simple to make.

First, drill a hole all the way through one of the plates. Too small a hole can fill with weld metal too fast, without the bottom piece getting hot enough to fuse with the filler metal. Be careful that you drill a large enough hole to allow an adequate amount of heat input into the bottom piece so you will have adequate fusion between the two pieces.

You can also drill a series of holes in a line or at four corners, depending on the design of your part. Then, butt the surface or edge of another piece against the hole and fill the hole with molten filler metal. Even a tee joint can be plug welded if the edge of the top piece is placed directly over a row of holes drilled down the center of the base plate. Slot welds are the same as plug welds, except the hole is elongated to allow more surface area to be joined.

MULTI-PASS WELDS

Thicker sections require multiple passes to fill completely. The first pass in a joint is called the root pass and is typically the most difficult to make successfully. The next pass is called a fill pass, which fills the joint almost up to the final size. The final pass is called the cover pass and determines the profile of your weld. Some industrial processes, such as SAW, are designed to deposit large amounts of filler metal in one pass. In a small shop, these same joints require multi-pass welds. Take the time to consider how to go about it. Using smaller-diameter electrodes (wires and welding rods) will require more passes to complete the weld. The more passes it takes to complete a weld, the more heat is put into the metal. This results in increased distortion, residual/internal stress, and a possible decrease in the mechanical properties of the base metal. Larger-diameter electrodes work better for multi-pass welds, but require a welding machine with a greater output capacity to handle their size.

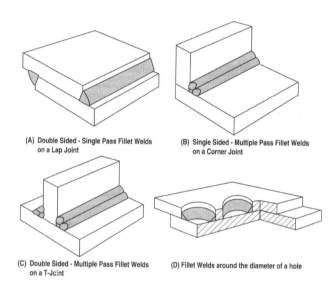

(A) Double Sided - Single Pass Fillet Welds on a Lap Joint

(B) Single Sided - Multiple Pass Fillet Welds on a Corner Joint

(C) Double Sided - Multiple Pass Fillet Welds on a T-Joint

(D) Fillet Welds around the diameter of a hole

Welding Inspection Technology, Fourth Edition, Figure 4.16, Reproduced with permission from the American Welding Society (AWS), Miami, FL USA

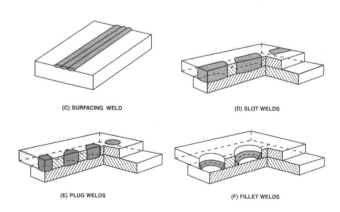

(C) SURFACING WELD

(D) SLOT WELDS

(E) PLUG WELDS

(F) FILLET WELDS

AWS A3.0:2001, Figure 15, Reproduced with permission from the American Welding Society (AWS), Miami, FL USA

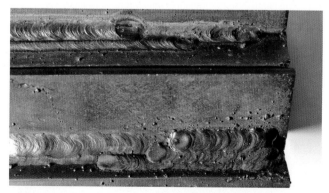

Multiple passes are used to build up a fillet-weld size. The joint at the top is a tee joint with three passes, below is a tee joint with six passes, showing the placement of passes four, five, and six. Each set of passes is made in a progression from the base plate to the top plate. *Monte Swann*

Pictured on the left is a multi-pass groove weld; notice the final pass was made by weaving the electrode back and forth. The distinctive look of the finished weave bead is shown on the right. *Monte Swann*

The cover pass on this horizontal joint was made with a slanted weave motion. *Monte Swann*

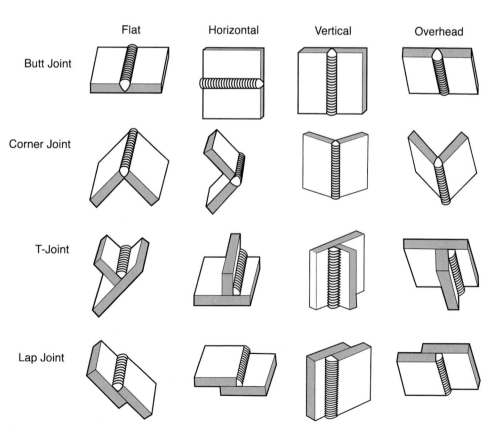

The orientation of the ripples in a weld bead will indicate the direction of welding. In this example, two beads were used. Both beads began in the bottom left corner. One bead continues along the short side and turns 90 degrees to finish at the end of the long side. The other bead continues along the long side and turns 90 degrees to finish at the end of the short side. Both beads end in the top right corner. Notice the crescent-shaped ripples point in the opposite direction of the weld progression. *Monte Swann*

STRINGER BEADS AND WEAVE BEADS

A typical weld bead is called a stringer bead. All of the exercises in this book use stringer beads. Although you may use a small circular or side-to-side motion during welding, this doesn't mean you are making a weave bead. Weave beads are usually no more than triple the width of a stringer bead and are used to deposit larger amounts of filler material into a joint in one pass. Weave beads are made with a side-to-side or zigzag motion. They require a little more skill to manipulate the puddle correctly and achieve adequate, even fusion without any discontinuities. Once you are able to consistently lay down good stringer beads, try weaving.

JOINT DESIGNS

A weld joint describes how two pieces of metal are placed in relation to each other. The type of joint to be welded depends on who designs the product. If you have designed the product or are building your own machine, make some educated decisions about joint design. Know your options and some of the unique aspects of welding each type of joint.

Corner joints

There are three types of corner joints. In open-corner joints the two pieces of metal meet corner to corner at the edges, without overlapping. This type of corner allows for a full-penetration weld without beveling the pieces. But, because of the small cross-section in the root of the joint where the two corners come together, it is a lot easier to burn through. In a half-open corner, one of the edges overlaps the other, about half the thickness of the metal. In a closed-corner joint, one edge fully overlaps the full thickness of the other piece, creating an L shape. An open-corner joint gives the final weld a rounded appearance. A closed-corner joint has a square edge, but may require beveling one piece to achieve adequate penetration, similar to a square butt joint. Welding the outside of a corner joint requires less heat than other types of joints.

Joint designs and welding positions. *Rob Lindgren*

Here is an open-corner joint at a 45-degree angle with a root opening to allow for complete joint penetration. *Monte Swann*

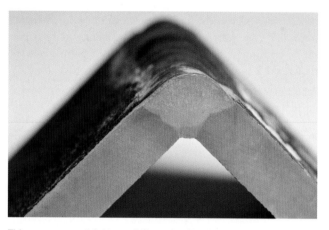

This open-corner joint has a full-penetration weld made with three passes using an E-7018 electrode. *Monte Swann*

Butt joints

The decision to prepare a butt joint by beveling or leaving a root opening (gap) between the plates will depend upon the following: metal thickness, if the joint is to be welded on both sides or one side, and the amount of penetration required. Every circumstance will be different, so by practicing welding on a variety of butt joints and testing/inspecting them, you will get to know how the geometry of the joint affects its practical application. Square butt joints have no bevel or edge preparation, but are frequently used with a root opening.

Groove welds on butt joints made with thinner metals 1/8 inch or less require a similar amount of heat to running a bead on a plate. Too much heat will cause burn-through (excessive penetration). Use less heat, faster travel speeds, and a smaller root opening (gap between plates) individually or in combination to eliminate burn-through. Be careful not to overcompensate and underweld your work. An equal amount of heat should be directed at each edge of a butt joint, unless the pieces are not the same thickness, in which case more heat should be directed at the thicker of the two. If the pieces are

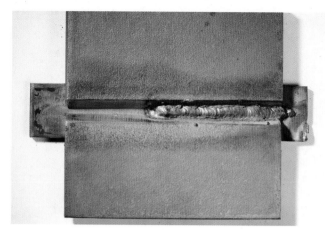

This single V-groove weld has bevels with feathered edges and a backup bar. The backup bar extends beyond the length of the joint and can be used as run-on and run-off tabs. A partial first pass has been made in the joint. Once the first pass is completed, the joint needs to be cleaned carefully to remove all of the slag. *Monte Swann*

the same thickness, but more heat is put to one edge because of improper welding angles, there will be a lack of fusion on the colder side and less penetration into the joint.

Lap joint

On the same thickness of metal, a lap joint requires more heat than a butt joint or corner joint. That additional heat is needed to get good fusion into the base plate. The edge of the piece on top of the base plate becomes one leg of the fillet weld. That leg cannot be any larger than the thickness of metal. If an equal amount of heat is directed at the base plate and the upper plate edge, the edge will always melt away before the base plate gets hot enough to liquefy. When welding lap joints, direct a majority of the heat toward the base plate and let the puddle wash up on the edge.

If this multi-pass lap joint was welded in its current position, the welds on the top left side would have been made in the horizontal position. The welds on the bottom right side would be made in the overhead position. *Monte Swann*

Tee joint

Joints where the pieces come together at right angles to each other are called tee joints. A tee joint has welding surface areas (fusion faces) in close proximity to each other and more area for the heat to dissipate to than any other basic joint. That is why tee joints require more heat for proper fusion and good welding than the other joints. Turn up your heat when applying a fillet weld to a tee joint.

There are joints that need as much heat as a tee joint, such as the root pass of a V-groove weld with a ¼-inch root opening and backing plate. Welding this first pass requires more heat for the same reason a tee joint does—the close proximity of multiple fusion faces on the joint.

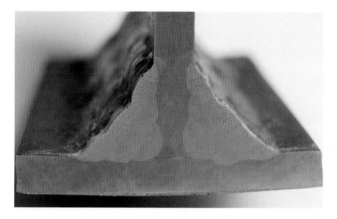

This is a multi-pass fillet weld applied to both sides of a tee joint. The weld has been cut, polished, and etched to show the difference between base metal, filler metal, and the fusion zones in between. *Monte Swann*

Tubular connections are made in a T, K, or Y shape. The final pass of this Y connection was made with multiple starts and stops and with a weave progression to cover the entire joint. *Monte Swann*

Edge joint

This uncommon joint is used to connect two very thin pieces of metal by bending the ends of each piece into an L, hooking the two Ls together, and welding along the joint. This is the least common type of joint, but has specific applications in manufacturing.

WELDING POSITIONS
Flat

Anytime you are learning to weld, are welding a new kind of material, or using a new welding process, start by welding in the flat position. This is the easiest position to weld in, and you can be comfortable and steady. There is one odd thing to consider about fillet welds made in the flat position. If a tee joint is placed flat on the table, it would actually be in the horizontal position. When tipped at a 45-degree angle, that same tee joint is in the flat position. Think of capping the ends of that tee joint and filling it with water. The tee joint creates a trough, and if held at a 45-degree angle, the water stays in. When the tee joint is rotated and is flat on the table, the water will spill out of that trough. Instead of water, think of liquid metal in the molten pool of your weld. The same principles of gravity apply. The face of a tee joint in the flat position points straight up.

A bead on a plate is technically a surfacing weld because there is no joint involved. The spacers underneath the piece help control the amount of heat in the base metal. If it were placed directly on the table, the large metal top would act as a heat sink drawing the heat away from the small piece of metal. *Monte Swann*

This is a tee joint in the flat position at a 45-degree angle from the table. *Monte Swann*

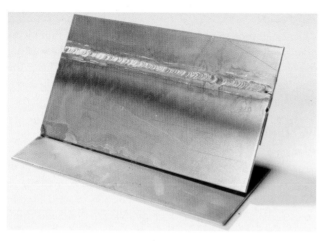

Here's a lap joint in the flat position with a single-pass fillet weld made with the GTAW process on 14-gauge CRS sheet. *Monte Swann*

Horizontal

When welding joints in the horizontal position, gravity becomes a factor. Good welding technique and proper welding angles will help compensate for the effects of gravity on the molten pool. Horizontal welds are more difficult than flat-position welds, but should be practiced before attempting welds out of position (vertical and overhead). However, horizontal welds generally keep your back straight (causing less fatigue) and the smoke and fumes out of your breathing zone.

Vertical

The weld characteristics applied to vertical joints can change drastically depending on which direction you travel. Downhill is welding in a downward progression and is easier because the weld pool is following the pull of gravity. Problems can occur when too much filler material is added, overflowing the top of the molten pool and only sitting on the surface of the metal. This overlapping of the filler metal over the base metal can be avoided by closely watching the molten pool for proper fusion.

Downhill welding will provide less penetration into the base metal. Shallow penetration characteristics are useful for thinner materials ⅛ inch or less, because you would likely burn through them welding uphill. But on thicker pieces (over ⅛ inch), uphill welding is required for adequate penetration. Uphill welding is more difficult because you are working against gravity, traveling in the opposite direction of its pull.

Vertical welds are more sensitive to the amount of heat input. If the molten pool is allowed to grow too big or too fluid, the liquid metal will spill out of the puddle and run down the face of the joint. To weld successfully in the vertical position, good welding technique and the correct amount of heat (usually a little less than what is used for the other welding positions) becomes more important, since there is less of a "fudge factor" than there is with welding in-position (flat and horizontal).

Overhead

Welding overhead is made possible by the fact that liquid metal has a fair amount of surface tension and will stay in the molten puddle as long as it is not allowed to become too big and form a droplet. The amount of heat required to weld overhead is similar to the flat position. Set the welding machine controls and make test welds in the flat position so you will be close to what is needed to weld overhead.

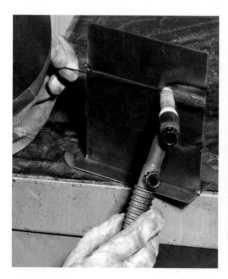

A square butt joint (without an edge preparation like a bevel) open-root groove weld is being applied in the horizontal position. *Monte Swann*

This tee joint is being welded in the horizontal position with the wirefeed gun tilted down slightly to compensate for the effects of gravity on the molten weld pool. *Monte Swann*

Here is a lap joint being welded in the horizontal position with the GTAW torch angled to direct the proper amount of heat to achieve proper fusion along the lower piece. *Monte Swann*

To be in the overhead position, the joint does not need to be head height. An upside-down weld joint 6 inches from the ground is still in the overhead position. What is critical is the orientation of the joint; the welder then works around it. *Monte Swann*

DISCONTINUITIES AND DEFECTS

The term discontinuity simply means something wrong with the weld. If enough discontinuities are present, or one is present that is very serious, it will make the weld too weak for the load requirements and stresses. The weld is then considered defective and the weld will likely fail.

Welds can also come apart because of poor design, overloading, and fatigue. Most discontinuities result from using incorrect materials and welding practices for the job, improper machine settings, or poor welding technique. Welding codes, such as the AWS D1.1 structural steel code, will have limits on the types, sizes, and quantities of various discontinuities. In fabricating and welding your own projects, you will have to decide if the welds are defective, which could be a matter of life or death and everything in between.

If the discontinuities need to be fixed, the weld metal must be removed by grinding, gouging, or cutting down to sound material (good metal). Rewelding over the top of slag inclusions and cold welds with incomplete fusion will most likely not fix the problem. Making more weld passes than necessary and overfilling a joint will not make the joint stronger. Instead, overwelding creates stress risers, causes excessive distortion, and weakens the base metal.

NOTCH EFFECT

A notch in a part concentrates stresses in that area, which can lead to premature failure. For an example, use any filler rod or paperclip and bend it back and forth until it breaks. Now take that same piece of metal and put a notch in it with a chipping hammer or chisel. Now bend it back and forth; it will break in just a few bends. This is the notch effect in action. The notch shape you put into the structure of that round piece of metal created a place for the stresses to concentrate, thus making it easier to break.

POROSITY

As its name implies, porosity is a void or hole in the weld metal, like swiss cheese or a sponge. Although the individual pores can be very small, porosity can be visible on the surface or be subsurface and not visible. Porosity is not as serious a discontinuity as others.

Porosity is caused by:

1. Atmospheric contamination due to a lack of a shielding atmosphere in the form of a shielding gas or flux, allowing oxygen, nitrogen, and hydrogen to be absorbed into the molten weld pool.
2. Overheating the weld metal and boiling the molten puddle. A normal amount of heat produces a normal amount of gases in the molten pool, which can escape before the puddle solidifies. Overheating the molten pool produces a larger volume of gases, which are unable to escape and are trapped as the metal solidifies.
3. Contaminants like rust, paint, scale oil, and dirt on the weld area of the base metal.
4. High levels of iron oxide in the base metal. When the base metal is melted, iron oxide combines with the carbon in steel to produce carbon monoxide/dioxide.

This short-circuit GMAW weld was made without any shielding gas. A large amount of porosity with very small pores is a discontinuity throughout the entire weld. *Monte Swann*

The pits in this weld are examples of a large-sized porosity caused by mill scale contaminants on the tubing picked up in the admixture of the weld. *Monte Swann*

Prevent porosity by using the correct shielding gas flow rates, proper welding technique, and correct electrode or filler rod. Also, do not suddenly withdraw your torch, electrode, or gun from the end of your weld. Clean your base metal joint thoroughly and use the proper amount of heat when welding.

INCLUSIONS

Inclusions are voids or holes in a weld filled with either slag or tungsten. Slag is the byproduct of the flux used in the SMAW and FCAW welding electrodes. Tungsten is the electrode used in the GTAW process. It is not possible to have inclusions with the GMAW and gas welding processes, which do not use flux or tungsten.

Although not as serious as other welding discontinuities, inclusions do not bond with the surrounding weld metal and ultimately weaken a joint. In SMAW and FCAW, slag inclusions are the result of low heat input, improper welding technique, and slag not being thoroughly removed by chipping, wire wheels, and picks before additional weld passes are made on top.

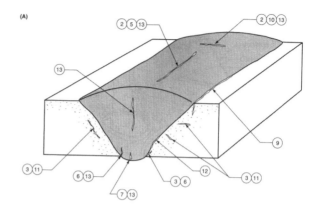

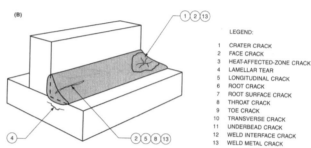

LEGEND:

1 CRATER CRACK
2 FACE CRACK
3 HEAT-AFFECTED-ZONE CRACK
4 LAMELLAR TEAR
5 LONGITUDINAL CRACK
6 ROOT CRACK
7 ROOT SURFACE CRACK
8 THROAT CRACK
9 TOE CRACK
10 TRANSVERSE CRACK
11 UNDERBEAD CRACK
12 WELD INTERFACE CRACK
13 WELD METAL CRACK

AWS A3.0:2001, Figure 33, Reproduced with permission from the American Welding Society (AWS), Miami, FL USA

This weld shows a lack of fusion, with the voids filled in with slag (slag inclusions). *Monte Swann*

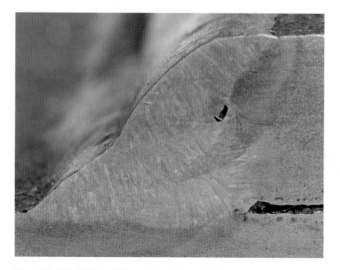

The pit in this fillet weld is a sub-surface slag inclusion between the first and third pass. *Monte Swann*

CRACKS

Any crack in a weld is a very serious discontinuity and usually will result in joints coming apart under any kind of load. Under most welding codes, cracks are never acceptable. Once a crack begins, it will travel down the entire length of the weld, making the joint come apart like a zipper. Micro-cracks, which are too small to see with the naked eye, can grow larger under applied loads. The location of a crack can tell you a lot about what caused it and how to avoid it in the future.

Segregation-induced cracking

Cracks induced by impurities are referred to as hot cracks and usually appear down the center of a weld bead. Some base metals have a large amount of impurities in them, such as phosphorous, zinc, copper, and sulfur. As the base metal is welded, these melted impurities mix with the filler metal in the weld pool (admixture), which has no immediate effect until the molten metal begins to solidify. Because the impurities have a lower melting point, they are the last to solidify. Since the weld bead solidifies adjacent to the base metal first (along the fusion faces) and last in the center, these melted impurities are forced to move to the center of the weld bead. The solidifying metal is forced apart by the still molten impurities, and the metal cracks.

Note: the same thing can happen when braze welding, when the base metal is overheated. If the bronze filler metal is allowed to flow into molten steel, the steel will solidify around the bronze and crack apart.

If you cannot get around the contaminants in the base metal, try reducing the admixture pickup of impurities in the base metal by:

1. Limiting the amount of penetration and depth of fusion into a joint.
2. Using large-diameter electrodes in combination with amperage and voltage setting on the lower end of the working range.
3. Keeping the width-to-depth ratios of single pass weld beads at either 1:1 or 1.4:1. Welds should be as wide as, or wider than, they are deep.

Aluminum cracking

In 6000 series aluminum, cracking occurs because magnesium silicide (around 1.0 percent alloyed in the base metal) becomes concentrated in the molten pool. When an aluminum filler metal is put into the admixture of the weld, it dilutes the magnesium silicide and prevents the crack from occurring. That is why a filler metal must always be used when welding this type of aluminum.

Weld profiles

Centerline cracks are more likely to occur in fillet welds with a concave profile. With this type of profile, the tensile strain in the weld shrinkage puts more stress on the solidifying metal. Fillet welds with flush or convex contours can be achieved:

1. In GMAW and FCAW, decreasing the arc voltage to return the bead shape to a convex profile.
2. By slower travel speeds, which allow more filler metal to enter the joint.
3. On vertical joints, welding uphill instead of downhill. Avoid excessive convexity to avoid the notch effect.

HAZ cracks

This type of crack occurs immediately adjacent to the weld bead either on the surface or subsurface. HAZ cracking is also called under-bead cracking or toe cracking, describing the location of the crack. Other names include delayed cracking or cold cracking because the cracks occur when the steel is cooled below 400 degrees F. On some grades of steel, cracks can show up as long as 48 hours after welding. The three conditions that cause HAZ cracks are:

1. *Presence of too much hydrogen*
   HAZ cracking is also known as hydrogen-assisted cracking, which describes one of the root causes for this type of crack. Hydrogen atoms are very small, so small that they can move between the grain boundaries of solid steel. To see this happen for yourself, make a weld on a piece of steel and let it air cool to room temperature. Get a clear glass baking dish and fill it with glycerin, which is found at any drugstore. Place the weld into the dish and watch. The line of tiny bubbles down the length of the bead is the hydrogen escaping from the weld metal. For comparison, make two SMAW beads, one with E6010 and one with E7018. There will be far fewer bubbles coming out of the E7018 bead because that electrode is designed to minimize the amount of hydrogen in the weld.

   Hydrogen comes from organic materials like water, rust, paint, dirt, grease, and oil. Hydrogen can be present in or on the base metal, electrode, flux, or shielding gas and in the atmosphere. When metal is liquefied, the molten pool has the ability to absorb and dissolve a large amount of hydrogen. As the metal solidifies, hydrogen can become trapped in the weld bead and heat affected zone. As excess hydrogen is trapped in pores between the grains of the base metal, a tremendous amount of pressure is created. Mild steels and lower-strength steels are ductile enough to compensate for this pressure and not crack. This is not the case with steels of high hardness and high strength. Using an electrode from the low-hydrogen group like E7018 and keeping the base metal joint clean and dry will greatly reduce the negative effects of hydrogen.

2. *Base metal sensitive to cracking*
   Some metals have a sensitive grain structure (microstructure) making them more susceptible to cracking. These include alloy steels (high strength), medium- to high-carbon steels (tool steels and heat-treatable steels), and all types of cast iron.

3. *Enough shrinkage stresses, residual stresses, or an applied load*
   Shrinkage stresses and residual stresses are closely tied to the rate of cooling. The quench effect increases both of these types of stresses because the rate of cooling determines the HAZ properties of the finished welds. The quench effect can result from:

   - Low-heat input welding procedures
   - Thick sections and large pieces of metal
   - Cold base-metal temperatures
   - Quenching hot metals in oil or water

The best way to reduce the quench effect is by preheating the base metal before welding. By preheating to 200–500 degrees F (depending on the type of metal) it greatly reduces the cooling rate. A slower cooling rate allows more time for the hydrogen to diffuse back into the base metal and create favorable HAZ properties to form, preventing cracks.

Measure your preheat temperature at least 3 inches away from the weld joint for metal up to 3 inches thick. Always preheat any metal up to room temperature, about 50 to 70 degrees F before welding.

Usually, preheating and controlling the presence of hydrogen is enough to prevent cracks. If it is not, post-weld heating can be performed between 400 and 450 degrees F held for approximately one hour for every inch of base-metal thickness. On small parts, it is always a good idea to slow down the cooling rate after welding by burying welded pieces in kitty litter, dry sand, or wrapping them in a fire blanket.

Transverse cracks

A less common type of crack known as transverse cracks shows up across the weld, 90 degrees to the length of the bead. Transverse cracks can be caused by too much hydrogen in the weld bead. The second cause is using a filler metal with a much higher tensile strength than the base metal. For example, using an E11018 electrode as your filler metal (110,000 psi tensile strength) to weld A36 steel (36,000 psi minimum tensile strength).

Crater cracks

As a general rule, discontinuities are most likely to occur at the beginning and at the end of weld beads. The end of a weld bead is sometimes left below flush, having a convex appearance that is referred to as a crater. Craters or porosity left at the end of a weld bead are more likely to cause cracks than craters that are filled to a flush or convex profile. Crater cracks at the end of a bead are easily preventable. If left unchecked, these cracks are likely to travel down the entire length of the weld, when any type of load is applied.

Crater cracks are caused by the following:

- The welding gun or torch is abruptly removed at the end of a weld before the molten pool has enough time to solidify.
- Not enough filler metal is put into the end crater of a weld, giving it a concave appearance.
- Moisture in the shielding gas.
- Contaminants on the base metal.

Fixing cracks

Before you fix a crack, you need to determine the extent of it, and to do that effectively, you need to use a dye-penetrant test. Dye penetrant is also used to detect leaks in welded seams of tanks and boilers. Penetrant testing materials are available at most welding suppliers. If you want to learn more about the process, read the material supplied with the kit that tells you how to correctly perform this non-destructive test.

Cracked weld metal will need to be removed either by grinding, gouging, or cutting back down to the root of the joint or sound metal 2 inches beyond each end of the crack. Before rewelding, allow for enough clearance when beveling and adjust your welding procedures and technique to prevent cracks from reoccurring. Holes can be drilled 2 inches from each end of a crack to aid in stopping it from spreading, however this doesn't work in all cases.

OVERLAP

When the molten weld pool becomes too large, it can spill over the joint area and onto the adjacent base metal. This is known as overlap. Horizontal welds are especially prone to overlap because the weld pool wants to drip and run down the length of the weld. The sharp transition between the base metal and filler metal from overlap creates a notch effect stress riser. Like a notch in a paperclip or filler rod that weakens the metal, sharp transitions between the weld bead and base metal are a place stresses will concentrate. These specific shapes or contours can weaken the joint and structure, causing it to break.

UNDERCUT

In welding, the heat of the flame or electric arc is designed to melt the base metal. The filler metal in a rod or electrode is designed to fill in that molten pool and reinforce the joint. Undercut results when not enough filler metal has been put back into the weld area. A small groove at the toe of the weld can easily be seen by aiming a flashlight across the base metal toward the weld. Undercut actually decreases the thickness of the base metal at that point.

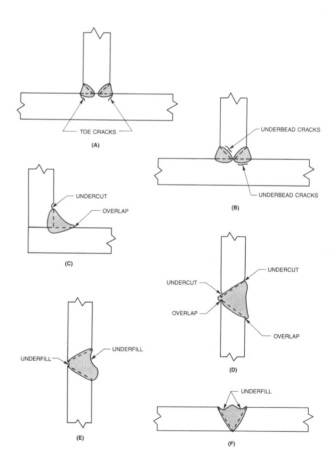

AWS A3.0:2001, Figure 32, Reproduced with permission from the American Welding Society (AWS), Miami, FL USA

Here you can see undercut along the toe of the bead due to an incorrect electrode angle and large pieces of spatter because of too long an arc length. *Monte Swann*

A small amount of undercut is usually not a concern on metals ⅛-inch thick or more. A small amount of undercut on thin materials, or a large amount on thicker materials, greatly reduces the cross-section of the base metal, making it less strong. Undercut also creates a notch condition in the joint, concentrating the stresses when a load is applied. Joints welded in the horizontal and vertical positions are more likely to have undercut.

Undercut is caused by:

1. Improper electrode angle for the type of joint and welding position.
2. Distance of torch or electrode being too far away from the base metal. The heat of the flame or arc spreads out the further away the torch or electrode is from the work. This will melt or cut away a large area, which cannot be filled in.
3. Improper welding technique. Fast, sudden movements, like stirring or shaking an electrode, does not give the filler metal enough time to deposit into the weld joint.

Severe undercut can be filled in by welding. This should only be done if necessary, since reheating the metal for additional welding results in increased residual stresses in the joint and may affect the mechanical properties of the base metal. Also, by rewelding you run the risk of creating more undercut along the toe of the new bead.

INCOMPLETE FUSION

Incomplete fusion is a very serious discontinuity. One of the basic principles of welding is the ability to fuse metals together. A lack of fusion results when a section of the base metal or part of the filler metal does not melt properly. Fusion will result only when the base metal and filler metal are both at the proper temperature. Groove welds and fillet welds can have areas of incomplete fusion at the fusion faces of the joint. The root of a fillet weld should have complete fusion in order to be full strength. Incomplete fusion can sometimes be seen

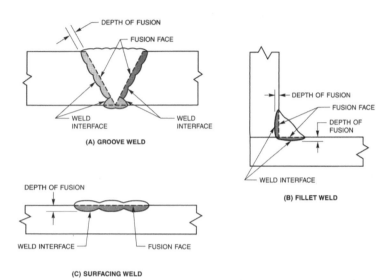

AWS A3.0:2001, Figure 30, Reproduced with permission from the American Welding Society (AWS), Miami, FL USA

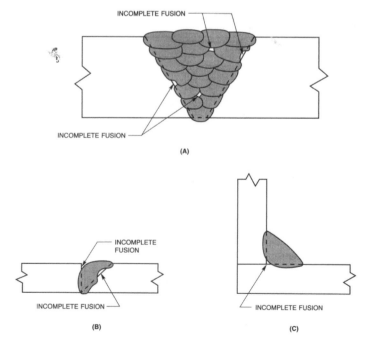

AWS A3.0:2001, Figure 29, Reproduced with permission from the American Welding Society (AWS), Miami, FL USA

on the surface of the joint where the arc did not melt the base metal or the filler metal is mostly sitting on top of the joint. In other cases, incomplete fusion can only be determined by testing the weld.

Incomplete fusion is caused by:

1. Not enough heat for the electrode, metal thickness, or joint design.
2. Incorrect electrode angle resulting in not enough heat in one part of the joint.
3. Erratic travel speed, skipping ahead, and not giving an adequate amount of time for fusion to take place over a certain area.
4. Allowing a buildup of slag or molten metal under the welding arc, interfering with its ability to adequately heat the base metal.

The fissures at the base of this tee joint are examples of incomplete fusion. *Monte Swann*

This weld shows complete fusion between the layers of surfacing welds. *Monte Swann*

5. Impurities on the base metal—like paint—which causes an erratic arc and mill scale and can take the heat away from the weld.

In the GMAW, FCAW, and SMAW welding processes, incomplete fusion can only be fixed by grinding out the area where fusion did not occur and carefully rewelding. Make the proper adjustments in heat and welding technique before rewelding. Reheating the metal is possible in gas welding and GTAW, blending the base and filler metals together.

INCOMPLETE PENETRATION

Since only groove welds have joint penetration, only groove welds can have incomplete penetration. A groove weld with complete joint penetration (CJP) will typically have filler metal fused completely with the base metal all the way through the center and on both sides of the joint. Look at the back of a joint welded from one side. If there is evidence that edge of the joint did not melt, and filler metal has not fused to the backside, then there is likely to be incomplete penetration. If the joint is welded from both sides, there is almost no way of telling visually if CJP has been achieved. The weld would need to be tested by X-ray, radiography, or ultrasonic methods (all non-destructive to the weld) or cut apart so the weld could be seen from the inside. The common causes of incomplete penetration are:

1. Not keeping the welding arc located on the leading edge of the weld pool. Allowing a large amount of slag or molten metal to flow in the arc gap can reduce the amount of fusion and penetration into the base metal.
2. Heat input is too low.
3. Bad joint design with shallow groove angles and/or small root openings.
4. Edges of a groove weld joint should always be beveled or have a root opening when grinding the weld face flush to the base metal. This will allow for adequate penetration.

Some welds require CJP for strength and load requirements. For welds that can be accessed from both sides, the side opposite the weld can be beveled by grinding or gouging down to the root of the previous weld and then finished. Welds that can be accessed from only one side, like pipes, will have to be cut apart and rewelded if CJP is required.

EXCESSIVE CONVEXITY/REINFORCEMENT

Sometimes discontinuity is obvious because the weld is extremely oversized. In other cases, weld profiles are measured with fillet weld gauges, V-wac gauges or other measuring devices to determine if too much filler metal was deposited. Only fillet welds can have excessive convexity and only groove welds can have excessive reinforcement. Both are a discontinuity for the following reasons:

Excessive penetration on this weld burned through the base metal. *Monte Swann*

Excessive penetration caused the base metal to sag on the backside, giving it a convex appearance. *Monte Swann*

A fillet weld on a tee joint with excessive convexity. *Monte Swann*

1. Transition at the toes of the weld is sharp, not smooth. This creates a notch condition in the joint. From an engineering standpoint, a sharp transition from the weld bead to the base metal is a stress riser—when a load is applied to the part it is more likely to break because of a notch. Some pieces of equipment used for mining operations are under a lot of stress and require a concave profile on all fillet welds to reduce or eliminate this notch condition
2. Excess filler metal is wasted.

Excessive convexity and reinforcement can be fixed by grinding or filing down the excess weld metal and shaping the weld bead for a smooth transition at the toes. Be careful not to gouge into the base metal creating another notch.

EXCESSIVE PENETRATION

Known as burn through, this discontinuity can be obvious when a hole appears, or more subtle in the form of a convex surface on the back side of the weld bead. Excessive penetration is usually the result of too much heat for the material thickness being welded and can be fixed by:

1. Reducing the torch tip or electrode size.
2. Reducing machine settings, such as amperage and voltage.
3. Controlling the overall heat put to the base metal.
4. Using faster travel speeds for welding.

Joint geometry can also be a factor, especially the amount of root opening and the shape of beveled edges. Having too much root opening is likely to cause excessive penetration. Use smaller root openings and chamfers on bevels to "carry" the heat of welding and avoid burn through. Be careful to leave enough root opening to avoid incomplete penetration.

ARC STRIKES AND SPATTER

Large pieces of spatter that fuse to the base metal and arc strikes (touching the welding electrode) outside the weld area can cause small heat affected zones to form. Some metals are so heat sensitive that they can become brittle and crack in that small zone. I have colleagues who have witnessed base-metal failure due to arc strikes. Spatter that can be easily removed is not a risk factor, but using anti-spatter products will not prevent large droplets from being a problem. Proper machine settings and welding technique will eliminate excessive spatter and arc strikes.

CHAPTER 6
OXYGEN, ACETYLENE, AND OTHER COMPRESSED GASES

Compressed gases are used in most manual welding processes. GMAW and GTAW welding processes use shielding gases to protect the molten puddle from being contaminated by the atmosphere. Some shielding gases are used in their pure form, such as carbon dioxide and argon, or mixed together, such as C25, a blend of 75 percent argon and 25 percent CO_2. Torch welding uses two main types of gas: oxygen and acetylene. This chapter will focus on setting up and handling the equipment used for gas welding, brazing, soldering, and cutting. This chapter also covers information and safety that applies to all compressed-gas cylinders, including those used for shielding gases.

OXYGEN

Oxygen by itself does not burn, but instead supports combustion. It is one of the three main parts of the fire triangle, the other two being fuel and a source of ignition. The atmosphere we breathe contains about 21 percent oxygen, enough for things to burn, but not enough to burn and weld steel. Oxygen in a compressed-gas cylinder is 99.5 percent pure. It is extremely important to know that fuel burns more rapidly and fires can ignite at a much lower kindling temperature in an oxygen-rich environment. Never mistake a cylinder of compressed oxygen for a cylinder of compressed air. Never use oxygen to run pneumatic tools, fill tires, or dust off your clothing or body. Oxygen can saturate the fibers of your clothing or hair and easily catch fire. Oxygen leaking into an enclosed space can have dire consequences.

Someone told me about an untrained employee of a fabrication company who happened to be the boss' son. He was working at the bottom of a large, tubular structure and decided to air it out during lunch by leaving the oxygen valve open and running. When he came back and lit his torch in that volatile atmosphere, he ignited the only fuel to be consumed by the fire: his clothes, hair, and skin. They pulled him out charred head to toe. Always remember you could be fuel for a fire under the right circumstances. Be very careful when handling and using oxygen. Even the electrical current from a light switch being turned on can start a fire in an oxygen-rich environment.

FUEL GASES

A wide variety of gases are used for brazing, cutting, and welding. Oxy fuel cutting (OFC) refers to the process of cutting metal using a fuel gas combined with oxygen. OFC-P indicates the fuel gas being utilized is propane. Oxy fuel

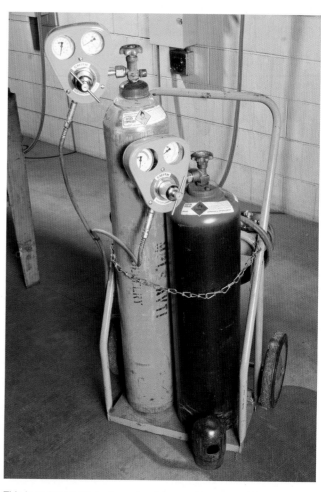

This is an industrial grade oxy/acetylene setup you could have at home, with a 150-cubic-foot acetylene cylinder and 125-cubic-foot oxygen cylinder. The cart makes it more portable. As a rule, always keep cylinder caps nearby so you can find them and use them when needed. *Monte Swann*

welding (OFW) is a generic name used for gas welding, but oxy/acetylene welding (OAW) is gas welding using acetylene as the fuel gas.

Fuel gases combined with atmospheric air (21 percent oxygen) burn at a lower temperature than when combined with pure oxygen. The flame temperatures given in the following section are for a neutral mixture of the fuel gas and high-purity oxygen. Combustion ratio refers to the amount of oxygen that is required for each fuel gas to burn in a neutral mixture. More information on the different types of flames will be covered later in this chapter. The following list contains information about each type of fuel gas and its uses.

Acetylene (C2H2)

Neutral flame temp: 5,600–6,000 degrees F

Combustion ratio: 2.5 parts oxygen to 1 part acetylene

Uses: Welding, brazing, cutting, heating

The oxy/acetylene flame is the hottest flame used for welding. This is the fuel gas of choice for welding ferrous metals, including steel, and is the most versatile in other applications such as brazing, cutting, and heating. Acetylene is more expensive than other welding fuel gases and the cylinders are heavier. Only a 3 percent solution of acetylene gas in the atmosphere can ignite. Acetylene gas is odorless, colorless, and tasteless but, like propane, a foul-odored chemical is added to the gas so when it is released you have a chance of detecting it. Acetylene gas should never be used in contact with copper or silver, as dangerous acetylides can form. Once formed, these acetylides can be ignited by a slight shock or heat. Other than welding/cutting tips and nozzles that are made of copper, any alloy containing more that 67 percent copper should not be used with acetylene. Use R-grade hoses with acetylene.

Propane (C8H)

Neutral flame temp: 4,580–5,000 degrees F

Combustion ratio: 5 parts oxygen to 1 part propane

Uses: Brazing, cutting, heating

Propane is popular because it is a commonly available fuel gas used primarily for cutting, heating, soldering, and light brazing. The regulator and torches used with acetylene will also work for propane. You will need to purchase a two-part cutting tip specifically designed for propane, which has the advantage of cutting through a wider range of metal thicknesses than tips for acetylene. The one drawback to using propane is the higher combustion ratio, which means a propane torch will use double the amount of oxygen. However, oxygen is inexpensive in relationship to the cost of fuel gases. The fuel gas is less expensive than acetylene and the cylinders will hold a larger volume of gas and are lighter weight. Rosebud-design propane heating torches can provide a large amount of heat in an economical way and would be a good choice for home workshops without acetylene. Use T-grade hoses with propane.

MPS or MAPP (C3H4)

Neutral flame temp: 5,200–5,400 degrees F

Combustion ratio: 4 parts oxygen to 1 part MPS

Uses: Brazing, heating

Methlacetylene-propadiene stabilized (MPS) gas is a commercially prepared blend of fuel gases (acetylene and propane) used for cutting, heating, and torch brazing. MPS provides a more even distribution of heat within the flame, requiring less manipulation to control heat input. MPS gas torches are popular because the gas costs less than acetylene and the cylinder contains a greater volume of fuel and is less heavy for the equivalent size. MPS gases are generally not used for welding ferrous metals due to the fuel-to-oxygen ratio required to raise the flame temperature and the resulting oxidizing characteristics.

Hydrogen

Hydrogen has a relatively low heat content, which makes it suitable for certain types of torch brazing and welding aluminum. To weld aluminum, an oxy/hydrogen torch is used, along with a cobalt blue lens in your welding goggles for a better view of the almost invisible flame. Either a paste flux and solid filler rod or a flux-core aluminum welding filler rod can be used. Hydrogen is not the only fuel gas that can be used to torch-weld aluminum; acetylene and propane will also work with the right equipment.

Other fuel gases include butane, propylene, and natural gas.

COMPRESSED-GAS CYLINDERS

Safety

The contents of a gas cylinder are under an extreme amount of pressure and are potentially very dangerous. There are safety rules related to all gas cylinders you should follow:

1. Always cap cylinders that are not in use. Cylinders should be capped when they are transported or stored. If an uncapped cylinder tips over and the valve stem breaks off, that cylinder will become a powerful rocket, taking out anything and anyone in the way.

2. Before removing a cylinder cap, be certain the cylinder is secured with a strap or chain. Even if the cylinder is empty, the first thing I do after removing a regulator is put the cap back on. Cylinder caps should be kept within easy reach. Always store compressed-gas cylinders in the upright position with the cap on.

3. Never weld on, or strike an arc on, a compressed-gas cylinder. Be aware that a cylinder is not grounded in any way during arc welding. If the cylinder becomes part of a welding circuit, and accidentally becomes arc-damaged due to electrical currents or careless handling of welding equipment, the cylinder could rupture at that moment or when it is refilled. Welding suppliers will not fill a compressed-gas cylinder that has any marks (from an arc) indicating an electrical current passed through it.

Before removing a cylinder cap, be certain that the cylinders are secured in some way. The chain on this cart prevents the cylinders from being easily knocked over or pulled over by the hoses. If the valve stem were to break off, the cylinder would become a rocket that could cause property damage or personal injury. *Monte Swann*

4. Never modify cylinders or use them for anything other than their intended purpose. Cylinders should not be used for rollers or supports of any kind. Cylinders should be protected from extreme weather conditions.
5. Mishandling the cylinders by dropping them or having objects dropped on them should be avoided.
6. Keep oil and grease away from all cylinders. The presence of any oil or grease on a cylinder containing oxygen is especially dangerous and could cause an explosion.
7. Do not transport cylinders in an enclosed vehicle. If the cylinder has a leak, it may fill the vehicle with gases hazardous to human health.
8. The only way to verify the contents of a cylinder is the label on the top of the bottle. If the label is missing, do not use the cylinder. Return it to the supplier.

Oxygen cylinders

Oxygen cylinders are hollow and seamless, made with solid steel that is drawn through a set of dies. The top is crimped on, creating a container capable of handling extremely high

pressure. Oxygen cylinders are fully charged at 2,200 psi at 70 degrees F. For comparison, think about how much pressure is needed to inflate your car tires (35 psi), or how much pressure is required to blow up one of those tires (100+ psi). Because the contents are under such enormous pressure, oxygen cylinders are equipped with a double-seat, high-pressure valve. The important thing to know is the valve fully seats in two positions, all the way open and all the way closed. If the cylinder is not fully opened when in use, oxygen can leak from the valve stem creating an oxygen-rich environment. Always open oxygen cylinders all the way. A safety device called a safety nut and disc is located on the side of the valve stem. It looks like a small tube with holes drilled in it and a brass nut as a cap. It is designed to burst at 3,000 psi, letting out all the contents of the cylinder. If there were a fire, or for any reason the pressure inside the cylinder became too great, this device will prevent the whole thing from exploding.

Acetylene cylinders

Acetylene cylinders are low-pressure cylinders, fully charged at 250 psi at 70 degrees F. The safety device on an acetylene cylinder is called a fusible plug, designed to melt at 212 degrees F. If the plug melts, it would release the extremely flammable gas inside, but is a better alternative to the whole thing becoming a bomb. The valve on an acetylene cylinder should be open at least ½ to ¾ of a turn to seat properly, but not more than 1½ turns. Otherwise, you can cause a restriction in the gas flow, which in turn can cause torch starvation. There is no need to open the cylinder valve all the way. In fact, you may want to open it only one full turn so if there is a problem, the valve can be shut off quickly. For the same reason, if your cylinder has a key handle for the valve, keep the key wrench with the cylinder.

Acetylene gas becomes unstable at pressures above 15 psi. In order to fully pressurize a cylinder, it is packed with a porous substance, like spray foam, and saturated with liquid acetone. This keeps the acetylene gas in a stable solution until

Opposite the regulator fitting nut on the valve stem is the safety nut and disc. The multiple holes are there so if the gas is released in an emergency, the cylinder will not take off in one direction. Cylinder threads are either fine or coarse. The threads on the top of this oxygen cylinder are fine and would require a cap with fine threads. *Monte Swann*

You can clearly see the acetylene cylinder information stamped on top, including its serial number and DOT specifications, a label indicating the type and flammability level of the gas, and the two fusible plugs on each side of the cylinder valve. Notice the top of the cylinder valve has lettering and arrows to indicate which direction is open and closed. *Monte Swann*

it is withdrawn from the cylinder. Because of the presence of liquid acetone, there are a few safety concerns when handling and using acetylene cylinders.

If you find an acetylene cylinder lying on its side, do not use it right away. If it has been lying horizontal for some time, the liquid acetone will have settled, filing the valve stem area. In order for the contents to settle, the cylinder should be place upright for six hours before using. Otherwise, liquid acetone can be drawn through the cylinder valve, damaging the regulator and torch and possibly causing a fire or explosion. For that same reason, an acetylene cylinder should never be drained all the way. After all the acetylene is gone, liquid acetone may be the next thing drawn from the cylinder. Consider an acetylene cylinder empty at about 10 psi. If for any reason you see liquid flames dripping out of the end of a torch, shut down the system immediately and fix the problem.

Types and sizes

Compressed-gas cylinders are sized by the cubic feet of gas they can hold. Typical sizes are 300 (largest), 251 (largest for oxygen), 125, 80, and 40 cubic feet. Cylinders are specific to the type of gases they contain. The size cylinder you need depends upon a number of factors. Before deciding which one to use, determine how often you will use it. Small cylinders are lightweight and convenient to carry and store, but they run out of gas quickly. Certain torch body attachments require a larger-capacity cylinder due to their gas flow requirements or withdraw rate in cubic feet per hour (CFH). The rule is, no more than 1/7 of a cylinder's contents should be withdrawn in one hour. See "torch body attachments" for more information on this. At a welding supplier, you will be able to see the different size cylinders for yourself and find a size right for your own shop.

Inspection and testing

In the United States, gas cylinders are designed and maintained in accordance with the U.S Department of Transportation (DOT). As part of their maintenance, cylinders undergo hydrostatic testing, where they are pressurized with water beyond their rated capacity. This testing is done every ten years on cylinders with a star stamped on the top, and every five years for ones without the star. Tests are performed to prove the cylinder is still functioning properly. If a cylinder fails a test, the DOT serial number is stamped over with XXXX. No supplier will fill a compressed cylinder with the DOT numbers X'd out. Also, leave refilling your cylinder to the experts. Filling one cylinder from another is dangerous and should not be attempted by anyone other than a recognized gas supplier.

Buying compressed-gas cylinders

Cylinders up to 40 cubic feet in size can be purchased, filled, and exchanged without any paperwork. For example, a common 40-pound propane cylinder is not tested and is not required to be capped when transported or stored. Larger cylinders can be purchased from a welding supplier and come with a title of ownership. This title is critical to proving your ownership of that cylinder; without it, suppliers are unlikely to refill it. Never buy a cylinder on the Internet or in an auction, or you may get stuck with a cylinder you can't get filled. Most welding suppliers will also cover the cost of testing the cylinder. CO_2 cylinders are tested every five years and other gas cylinders every ten. The only drawback to buying the cylinder is that you can get a full refund in the first six months to one year; after that the cylinder losses its value, usually 10 percent per year. Eventually, the cylinder cannot be returned for a refund, and will have to be sold back to a supplier at a lower price. None of this is a problem if you plan on keeping your cylinders for many years and take good care of them.

Most welding suppliers will not deliver to residential homes, so you will need to have a way to transport your cylinder safely. Remember: an enclosed vehicle is not safe. If the cylinder is laid horizontally, it must be capped. If transported vertically, it must be secured with either a cap or regulator attached. A welding supplier has an interest in making certain that you are leaving their facility without violating any DOT rules; the fines for violations run in the thousands of dollars. I've seen city trucks with uncapped cylinders driving down the highway at high speed and thought it is a disaster waiting to happen. Even though this may not be a direct violation of OSHA regulations, I think it is better to take extra care when transporting cylinders.

Cylinders can also be rented or leased. The only advantage to this is if a company needs a large quantity of cylinders for a short amount of time for doing specialty jobs.

This Smith regulator has two gauges. The one on the right, closest to the cylinder, is the high-pressure gauge, which indicates how much gas is left in the cylinder. The gauge on the left is the low-pressure gauge, which indicates the pressure in the gas line. Each gauge has two sets of numbers, one set is the English system in psi, and the other numbers on the gauge are metric in kPa (kilopascals). *Monte Swann*

EQUIPMENT FOR OXY/ACETYLENE OUTFITS
Color coding
In the United States, hoses, regulators, and sometimes one-way check valves and flashback arrestors are color coded: red for fuel gases and green for oxygen. In Europe and other parts of the world, the oxygen hoses may be blue or black, and orange is used for fuel gases. Remember, cylinder colors are not regulated. An acetylene cylinder may be painted any color including green, and an oxygen cylinder may be painted red. The only way to tell which gas is contained in a cylinder is to read the label.

Threads
Fuel gas components like hoses, regulators, check valves, and even the threads on the cylinder and torch body are left-hand threads. Left-hand threads are turned to the right to loosen and left to tighten. A notch is machined in the nut to indicate left-hand threads. Oxygen components have right-hand threads. For safety, different threads are used to prevent accidentally assembling the equipment improperly.

Regulators
1. Use regulators specifically designed for a compressed gas. Never use an oxygen regulator for acetylene or vise versa.
2. Gas welding regulators come in two main types: single stage and dual stage. Light-duty single stage regulators are cheaper but are intended for lower pressure applications, such as air-acetylene torches. They provide a less consistent and precise flow of gas and need to be constantly adjusted to maintain the correct working pressure. A set of heavy-duty single stage regulators may work well enough for applications requiring a less

precise gase flow. A two-stage or dual-stage regulator first reduces cylinder pressure (first stage) to a lower pressure (second stage, usually 200 psi for oxygen and 50 psi for acetylene) and then withdraw gas from the reducing chamber for pressurizing the lines. These regulators function consistently well when used in high-pressure applications. They do however cost up to three times as much as a single-stage regulator. No matter which type you choose, having a good set of industrial grade regulators can save you extra expense down the road if you decide to upgrade to larger cylinders and torches.
3. Regulators require specialized knowledge to repair. Never try to repair a regulator yourself. If you are ever unsure if a regulator is functioning properly, bring it to your local welding supplier. It could be that the regulator looks worn, has been damaged when accidentally dropped, or is cross-threaded onto a cylinder; could be the needle on the gauges is creeping up (not remaining at the set pressure when the needle valves are closed) or the pointer on the gauge doesn't drop all the way down to zero when the pressure is released. If anything is out of the ordinary, have your regulator bench tested. A qualified supplier will send it out to a company that specializes in fixing regulators used with high-pressure gases. Better to have them tell you it is junk rather than have one explode in your shop.
4. Keep oil and grease away from your regulators, since the presence of these substances on regulators can be explosive. If you handle regulators with care and only use them for their intended purpose, they should last a lifetime.
5. On oxygen and acetylene regulators, the line or hose pressure is regulated by a pressure screw. Be certain to back out this adjusting screw before opening the cylinder valve. If you don't do this, the pressure in the cylinder will hammer the internal diaphragm of the regulator and, at least, greatly reduce its working life. At worst, the impact can strip the threading of the pressure screw and send unwanted shrapnel flying at high speeds.

Weep holes are drilled in the bonnet of the regulator. These holes relieve any pressure built up in the regulator. If any gas is leaking out of the hole, this means the diaphragm in the regulator is damaged. *Monte Swann*

This cutaway of a single-stage regulator provides us with a clear view of the spring-loaded internal diaphragm of the regulator on which the pressure adjustment screw acts to increase or decrease the gas pressures in the lines. Always remember to back out the pressure-adjusting screw before opening the cylinder valves. *Monte Swann*

Hoses

New hoses come in a variety of diameters. Long lengths of small-diameter hoses can cause gas restrictions and will tend to kink. Use the shortest length of hose and correct diameter for your work area. The proper diameter to use will depend upon the size of torch, gas-flow requirements, and overall hose length. Hoses damaged or cut can be repaired with new ferrules. Use the correct-grade hose for your fuel gas and consult your welding supplier for information and help with your hoses.

If you have old hoses, be certain to check them carefully and leak-test them since they can become porous over time (see the section on checking for leaks). Oil and grease can soak into them and cause erosion.

On top of the gas hoses is a pair of goggles with a No. 5 shade lens and striker for lighting the torch. The blue box has the different-sized reamers used to clean welding and cutting tips. Having a set of tip cleaners is essential. The brass fittings in the lower-right corner are one-way check valves and a flashback arrestor. *Monte Swann*

Safety equipment: check valves/flashback arrestors

One-way check valves can be installed anywhere in the system and are typically placed at the connection between the hoses and torch body or the regulator and hoses. Some regulators have check valves built in. These valves prevent the reverse flow of gases and in turn prevent the mixing of gases in the lines and regulators. Oxygen and acetylene are a powerful and volatile combination, so mixed gases in the system can cause an explosion.

Flashback arrestors installed between the hoses and regulators will stop a flashback—mixed gases burning back into the system—from continuing to burn into the regulators and cylinders. If a flashback should occur, the check valves and flashback arrestors must be replaced.

Torches

There are many torch manufacturers and a variety of types available. For welding steel, an oxy/acetylene torch is used. Smith and Victor are the two major manufacturers of oxy/acetylene equipment. I've used Smith regulators with Victor torches, but either brand will work great. Purchasing regulators and torches in a set will be less expensive than buying equipment separately. Larger-sized torches for cutting and rosebuds for heating with propane come in one piece and attach directly to the gas hoses, not requiring a torch body.

Torch body, also called a torch handle or blow-pipe, can be used with many tips and attachments for welding, brazing, cutting, and heating. The needle valves on the torch body regulate the flow of gases to the tip or attachment being used. The needle valves are stainless-steel, ball-seat valves and open and close easily. Overtightening a needle valve can cause leaks and will wreck the delicate valve.

Welding tips come in a variety of sizes and can be used for welding or brazing. Victor welding tips come in sizes ranging from No. 000 to No. 12, with 000 being the smallest. I've used Nos. 0, 1, 2, and 3 tips for a wide range of applications.

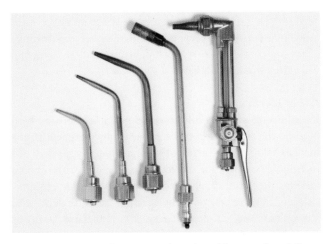

Victor torch body attachments are shown here. There are from left to right, No. 00-, 2-, and 3-size welding and brazing tips, a small rosebud heating tip, and torch cutting attachment. *Monte Swann*

Even three different tip sizes will handle a broad range of welding and brazing jobs. Whether you are buying new or have a torch already, know what size tips you own and find out if yours will meet the needs of jobs you plan to do. Welding thin steels or brazing may require a lower volume of heat and a smaller tip size. Welding or brazing thicker materials will require a larger tip size. Individual tips are worth purchasing to properly do a job. Tips and attachments have rubber O-rings to create a seal with the torch body. Using a wrench can crush the O-rings, so tips and attachments should only be tightened by hand. Missing O-rings and loose parts can cause flames to appear in places on the torch they shouldn't.

Cutting attachments are generally sized to the torch body being used. Cutting torches are standard sizes and are used with different sized tips. Each cutting tip is designed to cut a certain thickness range. Having the correct cutting tip for your attachment is one key to success. As with welding tips, find out what thickness ranges your torch and cutting tips are meant for and purchase additional ones when necessary.

Heating attachments, like rosebud torches, are great for heating large areas. They can be used for preheating metals, post-heating, limited heat treatment, and stress relieving. They are also a common tool used to heat metals for forming and sculpting. The most important thing to remember with these attachments is to have an adequate gas supply for the attachment being used. Either acetylene or propane can be use for heating. Propane may be a better choice, since it costs less and comes in cylinders that will hold a larger volume of gas.

$\frac{1}{7}$ rule

Each attachment, be it for welding, brazing, cutting, or heating, requires a certain supply of gas measured in a flow rate of CFH. Never use an attachment that requires more than $\frac{1}{7}$ the CFH of your total cylinder size. For example, if your oxy/acetylene heating attachment requires a 50 CFH flow rate of acetylene gas, multiply that number times seven (50 × 7 = 350 cubic feet). In order to safely use this attachment, you would need to have a total capacity that exceeds 350 cubic feet (this might require multiple cylinders). Remember this formula when you consider which attachments you will be using and what size cylinders you have.

Other torches

If you decide to use a different fuel gas, like propane, keep in mind the torches and tips are specifically designed for the fuel gas being used. Torches used for air-fuel gas are designed to aspirate (suck in) the correct amount of air for the chosen fuel gas pressure, usually set at 2 to 40 psi. For example, air-acetylene torches, or turbo torches, mix acetylene with atmospheric air (21 percent oxygen). The flame burns at a much lower temperature, which is great for brazing and soldering, but cannot be used for welding or cutting. MAPP (sometimes called MPS) propane torches are also commonly used in air-fuel applications, such as light brazing, soldering,

and heating. The plumbing, refrigeration, and electrical trades often use these torches in combination with small cylinders, which are easily transportable.

APPLICATIONS OF OXY/ACETYLENE

The oxy/acetylene torch is unique because of its high combustion intensity. Both the mixing of gases in combustion and welding or cutting with these gases is a chemical reaction. So, oxy/acetylene is a chemical welding and cutting process, and is the only combination of gases that has enough concentrated heat to weld steel.

Although the process has been around for the last 100 years, oxy/acetylene torches are commonly found on construction sites and in fabrication shops today. Even shops with sophisticated computer-controlled machinery will still probably have an old torch somewhere for maintenance and repairs. They may even use oxy/fuel gas torches in combination with CNC equipment as an economical way to cut through thick sections of steel. A major advantage of torches is that they are self-contained. There are no cords or plugs on an oxy/acetylene torch setup. No electricity is required, so it becomes a very portable option. The second advantage is its versatility. With the same torch you can weld, braze, cut, and heat just by changing tips, attachments, and gas pressures. The major limitation of the process is cutting. An oxy/acetylene torch can only be used for cutting plain carbon steels. Although thinner steels—⅛ inch or less—can be cut, they tend to bend and distort from the heat input. More information can be found in the cutting chapter. Welding can be done on a wide variety of metals, including aluminum and stainless steels, with oxy/acetylene. Welding these metals will require the correct torch, fluxes, welding procedures, and some amount of practice.

Setting up the torch

Your torch outfit should come with directions on its proper use. Be certain to read, understand, and follow all safety instructions provided by the manufacturer.

1. Chain or secure your cylinders to a fixed object or on a cylinder cart. Remove the protector cap and examine the inlet connection on the valve stem. Damaged threads are likely to ruin the regulator nut.

2. Crack open each cylinder valve, blowing out any dust or dirt particles that may have collected there. Opening and closing the valve quickly before installing the regulators will prevent dirt from clogging the system and dust from burning inside the oxygen regulator. Dust inside the regulator can cause an explosion due to friction and heat created by recompressing high-pressure oxygen.

3. Using a tight-fitting wrench, connect the regulators to the cylinders, being careful not to cross-thread the fittings or strip the nuts or threads. Don't use a pliers to tighten or loosen fittings. Next, be certain the

After chaining the cylinder and removing the cap, stand to the side of the cylinder and open and close the valve quickly. Do this to all cylinders before installing the regulator to blow out any dust or dirt in the cylinder valve. It is a good idea to stand to the side of the flow of gas, especially oxygen, which can saturate the fibers of your clothing causing them to ignite easily and burn rapidly from sparks or flames. *Monte Swann*

Be gentle and deliberate when installing a regulator. Carefully support the weight of regulators when installing them. Never try to force a regulator on or you can easily cross-thread the connection or strip the threads all together. If you encounter any resistance, inspect the cylinder inlet threads and regulator nut for damage or debris. *Monte Swann*

Fittings made of brass should never be overtightened and generally don't need a thread lubricant. After the fittings are snug, give the end of the wrench three light taps with your open hand. This will ensure that the leverage of the wrench doesn't damage the threads. *Monte Swann*

pressure-adjusting screw on the regulator is fully backed out, then slowly open the cylinder valve. Turn the adjusting screw clockwise until gas flows out. This allows dust and debris in that fitting or the regulator to be blown out. After a few seconds, back out the adjusting screw (turn counterclockwise) to stop the flow of gas.

4. If desired, install a flashback arrestor to the regulator outlet to stop a flashback or mixing of gases in the regulator.

5. Attach the hoses to the regulator or flashback arrestor: green to oxygen and red to acetylene. If you are using a flashback arrestor, use two close-fitting wrenches to tighten the connection. Use the same method of turning in and backing out the adjustment screws, one at a time, to blow out any dirt that may be lodged in the hoses.

6. Connect the torch to the hoses. One-way check valves can be installed at this location. The torch body (handle) will have markings indicating how the hoses are attached. Attach the red hose to the side marked FG and the green hose connected to the side marked OX.

7. Pressurize the system and check for leaks.

CHECKING FOR LEAKS

There are two ways to check for leaks in the system. The first way is to pressurize the system and shut off the cylinder valves. Look at the regulators and watch for the gauges to drop. If they do drop, there is a leak, although you may not be able to tell where it is. The second method is to pressurize the system and then apply a leak-detection solution. The solution can be purchased at a welding supplier, or you can mix a small amount of Ivory soap with water. A dish soap will also

Use two tight-fitting wrenches to install flashback arrestors between the hoses and regulators. Notice this flashback arrestor has a green stripe indicating it is to be used on the oxygen side of the system. *Monte Swann*

Don't Let This Happen to You

LOOSE HOSES CAUSED THIS FIRE

- Make certain all Oxy/Acetylene connections are tight
- Check for leaks with leak detection solution before lighting the torch

I made this sign after a close call in the welding lab. The loose connections leaked enough acetylene to be ignited by the flame of the torch. Fortunately, the person was able to shut off the needle valve immediately and was wearing protective clothing, preventing him from being severely burned. *Monte Swann*

work, but may be petroleum-based. Use only a few drops of soap in a spray bottle full of water. After spraying on fittings, if bubbles appear, you know you have a leak. If necessary, tighten that fitting with a wrench and retest; be careful not to overtighten connections made of soft materials, such as brass.

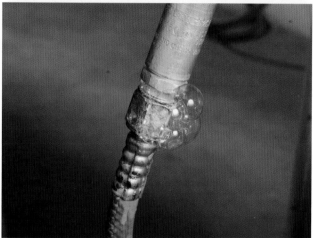

After attaching any components used with compressed gases, or if you think you have a leak somewhere in the system, apply a leak-detection solution. A few drops of dish soap mixed in water will work. Bubbles at the fitting indicate a leak. *Monte Swann*

If a regulator breaks apart or explodes, shrapnel can fly in any direction, but it is most likely to come out the front of the regulator. Stand to the side of any regulator when opening the cylinder valve. *Monte Swann*

If the connection is still leaking, bleed the system (see start-up and shutdown procedures) and examine the connecting nuts and screw threads for damage. Sometimes a small leak between the cylinder and regulator is expected because of the variety of cylinders constantly being rotated through welding gas suppliers. Return the cylinder if your regulator does not seat properly and the leak is severe. Keep in mind, even a small leak left unchecked over time can release a dangerous amount of compressed gas into an area.

OXY/FUEL STARTUP AND SHUTDOWN PROCEDURES

The following procedures are the steps I take in teaching how to safely use a torch. There are other ways of accomplishing this task, but I would strongly recommend following these steps. Not only will you set an accurate working pressure (pressure running through the lines when the needle valves are open), but you also purge the lines of any mixed gases before lighting the torch.

Pressurizing the system

- Make certain all valves are closed and back out the pressure-adjusting screws.
- Standing to the side of the regulator, open the fuel-gas cylinder valve ¾ to 1 turn.
- Open the fuel-gas valve on the torch body and turn the regulator pressure-adjusting screw clockwise to working pressure for the tip being used. This is read on the low-pressure gauge of the regulator. Remember: never use acetylene at pressures greater than 15 psi.
- Close the fuel-gas valve on the torch body.
- Standing to the side of the regulator, slowly crack and then open the oxygen cylinder valve all the way.
- Open the oxygen valve on the torch body and turn the regulator pressure-adjusting screw clockwise to working pressure for the tip being used. Close the oxygen valve on torch body.

The system is now pressurized and ready for torch ignition:

- Use only a striker to light a torch. Never use matches, lighters, hot metal, or another torch.
- Hold the striker to the side of the tip when igniting the fuel gas; do not cup it over the tip.
- Crack open the fuel-gas needle valve on the torch body and ignite the acetylene.
- Be certain to point the torch away from any flammable objects, gas cylinders, hoses, regulators, and people. Always keep track of where your torch is pointed.
- Adjust the acetylene needle valve on the torch body until you have the amount necessary for the volume of heat required for the job.
- Slowly open the oxygen valve and adjust to the required flame.

At this point, the oxygen cylinder has been opened and the high-pressure gauge reads 2,000 psi. We know the adjusting screw is backed out because the low-pressure gauge reads zero. *Monte Swann*

Hold onto the torch while turning the adjusting screw clockwise to pressurize the lines. This way the needle valve can be shut off easily. Be certain to watch the low-pressure gauge and stop when the gauge indicates the correct pressure for the tip being used. Be aware that the increments on the oxygen and acetylene low-pressure gauges are different. Make certain you are reading them correctly. *Monte Swann*

It is possible to back out the adjusting screw too far and have it come out. If this happens, inspect the screw threads for damage or debris and screw it back in. Since the internal diaphragm of the regulator is closed, no gas will leak out of this opening. *Monte Swann*

Shutting down the system or bleeding the lines
- Close the fuel gas valve on the torch body.
- Close the oxygen valve on the torch body.
- Close the valves on both of the gas cylinders.
- Open the fuel-gas valve on the torch body to bleed the gas from the line. Watch the regulator gauges and make certain both drop to zero.
- Open the oxygen valve on the torch body to bleed the oxygen line the same way.
- Back out both adjusting screws.
- Wrap up the hoses.

If you find that only the low-pressure gauge drops to zero and not the cylinder-pressure gauge, it could be because the pressure screws were backed out before the needle valves on the torch body were opened. If this happens:

- Leave the cylinder valves closed and reclose the needle valves.
- Turn the pressure-adjusting screw until there is a reading on the line-pressure gauge on the regulator.
- Re-open the needle valves, which should bleed the lines. Both gauges should then drop to zero.

Note: If the cylinder-pressure gauge rises after these steps, check to make certain the cylinder valve is fully closed. If it is, there may be a leak in the cylinder valve, which is a potential hazard.

OXY/ACETYLENE FLAMES
After you are comfortable with the equipment, the next thing to do is closely observe the different flames and be able to identify each of them. Be certain to wear a No. 5 shade so you can comfortably observe the different parts of the flame.

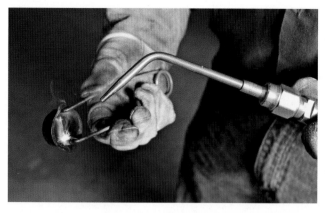

Hold the striker to the side when lighting a torch; never cup it over the end. Never use matches or lighters to light a torch. It is a good idea to keep these combustibles out of your pockets. *Monte Swann*

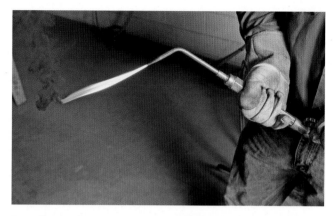

Open the acetylene valve slightly and ignite the torch; notice all the soot coming off the end of the orange flame. This greasy soot is carbon generated from burning pure acetylene. As acetylene is added the soot diminishes when the flame feathers out on the end. No matter the amount of acetylene you start with, the carbon soot will be eliminated once oxygen is added to the mixture. *Monte Swann*

Starting with a low amount of acetylene will produce a flame with a low volume of heat when mixed with oxygen. Having the acetylene set this low runs the risk of starving the torch tip, which could result in a flashback. A better option is to switch to a smaller-size welding tip to reduce the amount of heat. *Monte Swann*

If you open the acetylene needle valve too far, the flame will jump off of the torch tip. When the oxygen valve is opened at this point, the flame will extinguish itself. *Monte Swann*

This is how an acetylene flame should look before adding oxygen. The soot on the end has disappeared and the stem of the flame is next to the tip and feathers out nicely. *Monte Swann*

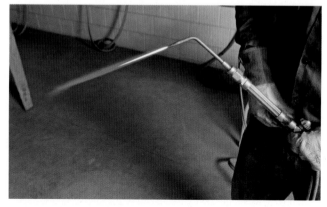

When oxygen is added to the acetylene flame it will begin to change shape and color. *Monte Swann*

After igniting and adjusting the fuel gas, slowly open the oxygen needle valve. If the torch makes a loud hissing sound and goes out, the mixture contains too much acetylene. Turn off the acetylene needle valve before relighting the torch. The flame should begin to change color when oxygen is added to the mixture. Continue to add oxygen until you can see three distinct parts of the flame: the cone, feather, and envelope. When you can see these three parts it is a carburizing flame. As you continue to add oxygen to the mixture of gases, the feather will become shorter. The point where the feather meets the cone and is no longer visible (all you can see is the cone and envelope) is a neutral flame. If you continue to add oxygen, the cone and envelope will become shorter, the envelope will turn a brighter purplish blue, and the hiss of the torch will become much louder. This is an oxidizing flame. It may be difficult to tell the difference between a neutral and an oxidizing flame since only the cone and envelope are visible on both. When in doubt, turn the oxygen down until the feather reappears in the carburizing flame. Then only add enough oxygen until the feather just disappears into the cone.

Applying the three different flames to a piece of steel will have different effects:

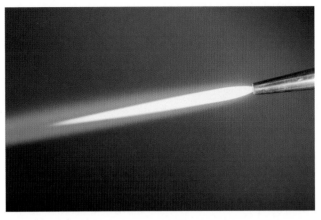

This is a carburizing flame. You can clearly see the feather and outer envelope, but the inner cone is not visible in this photograph. You will be able to see all the parts of the flame, including the inner cone, through a number No. 5 shade lens. *Monte Swann*

Carburizing flame: 5,700 degrees F

A carburizing flame has a large percentage of acetylene gas in the mixture. If you melt steel with the cone of this flame, the metal has a tendency to boil when the carbon from the flame is entering the molten metal. After the metal has cooled, look at the surface for pitting. Carburizing flames tend to make metals more brittle.

Neutral flame: 5,900 to 6,000 degrees F

A neutral mixture of oxygen and acetylene results in the gases burning off at an equal rate, completely burning all of the oxygen and fuel gas. When you melt steel with this flame, the molten puddle will be clean, clear, and will flow like syrup. This flame is used in most gas welding and cutting applications.

Both of these are neutral flames with different-sized inner cones. The larger cone has more acetylene in the mix and a greater volume of heat. The shorter cone has less acetylene in the mix and a smaller volume of heat. If the flame you are using is not hot enough, turn off the oxygen using the needle valve, turn up the acetylene, and remix for a neutral flame. If the flame is too hot, turn down the amount of acetylene. If the volume of heat coming from the flame is still not adequate, you will have to switch to a larger or smaller tip size. *Monte Swann*

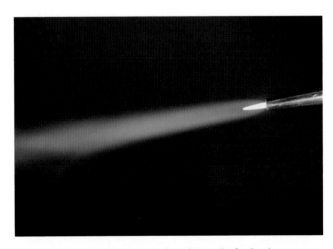

As you continue to add oxygen to the mixture, the feather becomes shorter. At the point in time when the feather meets the cone, you will have a neutral flame. Now the inner cone and envelope are visible. *Monte Swann*

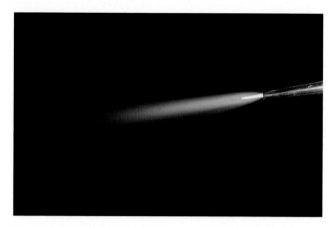

Too much oxygen in the mix produces an oxidizing flame. Notice the envelope is shorter, slightly purple in color, and less bright than before. *Monte Swann*

Oxidizing flame: 6,300 degrees F

An oxidizing flame is created by a high percentage of oxygen in the gas mixture. Putting this type of flame to a piece of steel and allowing the cone to touch the metal will result in lots of sparks and a white foam or scum on the surface. After this piece cools, the metal will look silver and shiny around the part that melted. Typically, an oxidizing flame has no useful purpose in welding or brazing.

BACKFIRE AND FLASHBACK

When a torch backfires, the flame goes out with a loud pop or series of pops that sounds like a string of firecrackers going off. This is the sound of the gases exploding inside the tip. Backfire often sends sparks of molten metal from your puddle flying everywhere. Typically, backfire is caused by:

- Overheating the tip.
- Operating a torch at lower pressures than required for the tip or attachment being used.
- Touching the tip against the work.
- An obstruction inside the tip.

See the section on torch popping in the gas welding chapter for more details on common causes and how to prevent this dangerous nuisance from happening.

Flashback is a more serious condition where the mixed gases are actually burning inside the torch tip. It is easy to identify because of the shrill hissing and squealing sound or lack of a visible flame. Sometimes a dirty tip will whistle, too, but the flame will still be visible and the tip will remain the same color. With an actual flashback, the tip will glow orange and begin building pressures in the torch that are higher than in the supply hoses. If allowed to continue, flashbacks will burn back into the system causing damage to equipment and possibly an explosion. If a flashback should occur shut the needle valves immediately and determine the cause. Flashbacks have many of the same causes as backfires, and can be prevented. Common causes include:

- Restrictions in the gas lines.
- Extreme lengths of small-diameter hoses.
- A torch tip placed directly on the metal.
- Improper gas pressures for the tip or attachment being used.
- Damaged or mishandled (overtightened) needle valves.
- Improper seating of the acetylene-cylinder valve.

All of these conditions can cause torch starvation, which in turn will overheat the torch causing a flashback.

CHAPTER 7
GAS WELDING

Acetylene was discovered in 1836, but it wasn't commonly used for gas welding until 1900 when a low-pressure torch was invented. Although gas welding is seldom used for joining metals in manufacturing, I believe it is the best process for the beginner to use to learn how to weld. Unlike SMAW, GMAW, or FCAW, gas welding takes place in slow motion—slow travel speeds with a clear view of the molten puddle. GTAW has the same educational advantages as gas welding, but can require more skill: manipulating the torch, filler rod, and amperage remote control (like a foot pedal) all at the same time.

In the last chapter we discussed the versatility of the oxy/acetylene torch. In this chapter we will utilize the torch to weld mild-steel sheet metal together. Sheet metals are easily welded using a torch, although there is more heat input to the base metal than with any of the arc-welding processes. Gas welding thicker metals requires a large-size tip that works at higher gas pressures. Thick base metals dissipate the heat, taking longer for a puddle to form, and the high gas pressures create turbulence in the molten pool. These factors limit the thickness of metals joined by gas welding to less than 1 inch. Metals over ¼-inch thick are usually joined with an arc-welding process instead of gas welding.

Low carbon and mild steels are easily welded with an oxy/acetylene torch.

Medium- to high-carbon steels, alloy steels, and the various types of cast irons are sensitive to the high amounts of heat required to create a molten puddle for welding. Great care must be taken to control the expansion and contraction forces associated with welding. Preheating and post-weld heat treatment will be required to maintain the mechanical properties and strength of these metals and to prevent cracking in the weld and heat-affected zones. Brazing and braze welding are in many cases a better option for joining these types of materials with a torch.

Copper and copper alloys, like brass, can be welded with an oxy/acetylene torch. A flux is required and a large welding tip size, since these materials quickly disperse the heat of the torch.

Aluminum can be welded with oxy/acetylene, but hydrogen is a better fuel-gas choice. Use the specific torch, lens, filler rod, and flux for gas welding or brazing aluminum.

Stainless steel can also be welded with the addition of a flux and smaller welding size tip to reduce the heat input.

For welding aluminum and stainless steels, I recommend using the GTAW process; copper or brass is easily brazed/soldered instead of welded.

FILLER RODS

Manganese and silicon content are very important in cleansing and deoxidizing the weld puddle, making the resulting weld less brittle and more ductile.

The two common rods (R) for gas welding (G) are:

1. RG 45 (45 ksi minimum tensile strength); copper coated to prevent rust
2. RG 60 (60 ksi minimum tensile strength); most commonly used filler rod; not copper coated

Other mild-steel filler rods can be used, such as ER 70S-2. See the GTAW chapter for more information.

Look for filler rods made by well-known manufacturers and information on the box about order or lot numbers and heat numbers, which identify that rod batch. This, along with an AWS designation, ensures the quality and traceability of the product. In structural welding and manufacturing, these standards are in place so if a weld fails, and a low-quality filler metal is to blame, the manufacturer can be held liable.

SETUP

Before you start any gas-welding project, take time to get set up for success. If you are having problems, your welding technique could need some work, in which case practice will help. If you are having trouble because your tools and equipment are not right for the job, or are being used

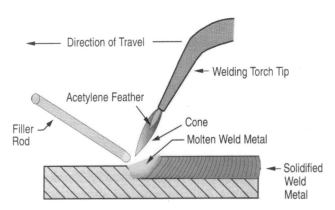

Welding Inspection Technology, Fourth Edition, Figure 3.35, Reproduced with permission from the American Welding Society (AWS), Miami, FL USA

Get set up with everything you will need before starting a weld. Point your torch in a safe direction, light it, and adjust the amount of acetylene. Then mix for a neutral flame. *Monte Swann*

incorrectly, it will not matter how good of a welder you are. It's like expecting someone to run a race with a broken ankle; no amount of talent can compensate for it. Be certain to get correctly set up for the job.

1. Inspect your work area for safety hazards of any kind. Inspect your torches for anything out of the ordinary, and wear the proper PPE.

2. Select the proper-size tip for the metal thickness to be welded. One of the common mistakes made during welding is not considering the thickness of metal. Metal thickness is the number one factor in selecting the proper size tip. Welding is about regulating the temperature of the base metal. Thicker sections require a greater volume of heat to weld successfully. A small tip will not provide the proper amount of heat to melt the base metals and filler rod. Using too large a tip on thinner sections will make the puddle too large and will probably end up burning a hole through the work piece. A tip chart should be supplied with any equipment you buy, can be provided by your local

welding supplier, or can be found at the manufacturer's website online. Tip charts give recommended tip sizes based only on metal thickness. Tip charts also give information on the size drill used to clean the tip, gas-pressure settings, and gas-flow requirements in CFH. Different manufacturers have different numbering systems and pressure requirements for their torches. Be certain to use the correct tip chart from the manufacturer.

3. Know and set the proper working pressure for the tip being used. Too little pressure can result in torch starvation, backfire, and flashback. Setting the gas pressure higher than what is required for the tip is not as bad as starving the tip. But excess oxygen pressure may have an oxidizing effect on a weld; and the more pressure acetylene gas is under, the more unstable it can become. It is best to use the recommended pressures for your size tip to eliminate this as a possible problem.

4. Filler-rod selection is determined by the type of base metal you will be welding. Select filler rods that most closely match the composition of the base metal or are best suited for welding that type of base metal. Every situation is different, so doing a little research on your part may be necessary. Ask a few questions at your welding supplier, or go online to find the information you need. You may get lucky by just using any old rod at hand, or even a wire coat hanger, but more than likely your welds will fail. Not to say that a coat hanger won't work. It will hold things together—like a broom rack, small sculpture, or something similar that is not under any type of load or stress. Some of the low-quality filler rods are much less expensive, but like coat hangers, are manufactured from materials

Be sure to measure your material thickness and select the proper tip size. This dial caliper will take quick and accurate measurements. *Monte Swann*

Tip cleaners are sold in sets and come with a small file for cleaning the face of the welding tip, but a small smooth file will do a better job. Don't try to force a reamer into the tip. Start by using a smaller size and work your way up until you find one that fits. The reamer sizes are stamped on the back of the box. *Monte Swann*

with lots of impurities or few cleansing and deoxidizing agents. As a rule, your filler rod should have mechanical properties equal to or greater than the base metal; it should be just as strong, if not stronger. Filler rods with AWS designations are a safe bet, because they are required to have specific mechanical properties and certain amounts of alloys and impurities.

5. Cleaning your torch tip, and keeping it clean, is easy to do. A dirty tip will change the shape of your flame and the direction of the heat. A dirty tip is one major cause of torch popping. Use a tip cleaner or file and proper-size drill bit in a pin vise. Keep the outside of your tip clean by wire brushing it to remove any soot, dirt, or slag. Use a smooth-tooth file to clean off the end of the tip until it is a bright copper color again, and then use the proper-size reamer to clean the inside of the tip. Start with a smaller-size reamer and work your way up to the size that easily will go in the end. Never force a reamer into the hole; it will likely break off inside the tip and will be difficult to remove.

WELDING TECHNIQUE

I don't tell students how to hold their torch because everyone is a different size and shape. Practice holding the torch first like a hammer and then like a pencil. Eventually you will hold the torch in different ways required by different jobs, but for now, start out with the most comfortable grip for you. Most important, relax your grip. You'll probably have a tight grip when you first start to weld—try to relax. Remember, none of the welding processes covered in this book require any physical force on your part in order to perform them correctly. Get comfortable. Position your arms

and body so you are relaxed and don't need to strain. The more you can support yourself when learning to weld, the better. Tired hands, arms, legs, and body make it difficult to concentrate on what you are doing, trying to make a good weld. Right-handed people typically hold the torch in their right hand and filler rod in their left; the opposite is true for lefties. There is no rule on trying both hands; some of the best welders are ambidextrous.

FOREHAND AND BACKHAND

When the torch is tilted at a travel angle, the direction of angle is an important factor in how your weld will turn out. In SMAW and GMAW this is known as push (forehand) and pull (backhand). Either way, forehand is traveling in the same direction the torch is pointed. Backhand is traveling in the opposite direction the torch is pointing. Use forehand technique for metals 1/8 inch or less in thickness and backhand for metals thicker than 1/8 inch.

DASH

There are four main factors in welding technique that determine if your weld will turn out right. They are distance, angle, speed, and heat (DASH). If you get these four things right, you will have a good weld every time. If you get one of these four factors wrong, your weld may or may not turn out. Two or more incorrect, and there is no hope of success. DASH factors change depending on the welding process used, thickness of metal, joint design, and welding position. For oxy/acetylene welding, there are many relationships among DASH factors to consider. Keep these relationships in mind when welding. The better you understand them, the better your chances for success.

Having your arms in the air like this makes keeping steady difficult. Steady yourself by using the table to brace your arms as much as possible and practice welding while sitting down. *Monte Swann*

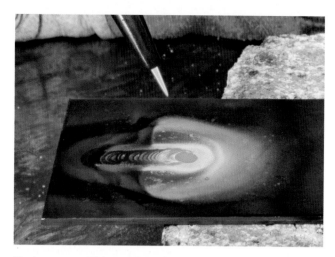

The inner cone of this torch is too far away from the base metal. Try to keep the cone no more than ⅛ inch away from the base metal. *Monte Swann*

This cone is too close to the base metal and is almost touching it. Notice that the turbulence of the flame is pushing the molten metal toward the edge of the puddle. *Monte Swann*

Distance

The tip of your inner cone should always be ⅛ inch from the metal. Keep this distance consistent throughout the weld.

Relationship to *heat*:

Only raise the torch and pull it away when removing heat. Holding the cone too close will overheat and contaminate the tip. Do not change distance in order to change the amount of heat when welding.

Angle

There are two angles to consider. Travel angle is the angle the torch is held in line with the weld axis (length). Work angle is the angle of the torch perpendicular to the weld axis (side to side). Both angles will change depending on the joint design and welding position, and direction of travel (forehand or backhand). See each exercise for the recommended angles.

Relationship to *heat*:

The input of heat into the base metal will change depending on your torch angle. Steep torch angles direct the flame's heat toward the work piece and into the base metal. Shallow angles (less than 45 degrees from the base metal) will cause more heat to bounce off and away from the work piece, putting less heat into the base metal.

Relationship to *speed:*

Travel faster with a steeper angle and travel slower with shallow angles.

Speed

A consistent travel speed will put a uniform amount of heat into the base metal along the weld joint and is the key to creating uniform weld beads. Look for a bead width from ¼ to ⁵⁄₁₆ inch. Slow travel speeds cause the puddle to spread out and become oversized and overheated, eventually boiling the molten metal and burning a hole through it. Fast travel speeds cause narrow beads and poor fusion into the base metal. Inconsistent speed will result in a bead with wavy edges or toes.

Relationship to *angle*:

Using the correct work and travel angles will help maintain a consistently sized molten weld puddle, making it possible to maintain a uniform travel speed.

Relationship to *heat:*

In general, faster travel speeds will apply less heat toward the work piece and into the base metal; slower travel speeds will apply more heat toward the work piece and into the base metal.

The corners of these two pieces of metal illustrate the importance of torch angle. Notice one corner is melted while the other one remains intact, even though both have been heated. Remember: The angle of your torch will direct the heat. *Monte Swann*

If your travel speed is too slow, eventually you will burn through your work piece. As an experiment, burn through a scrap piece of metal intentionally. This will tell you a lot about how the metal will behave when heated. *Monte Swann*

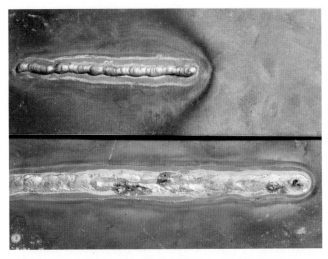

The amount of heat during welding will determine the profile of your weld bead. The weld bead pictured at the top was made without enough heat. You can see how it has a very convex appearance and is not fused at the toes (edges). Too much heat was used to weld the bottom bead. Notice the concave indentation through the middle part of the bead. Also notice the hole in the crater at the end of the weld. This was due to removing the torch suddenly, allowing oxygen to contaminate the molten puddle. *Monte Swann*

Heat

The volume of heat will depend upon the size of the tip being used and the amount of acetylene in the gas mixture. Increasing the acetylene flame will increase the volume of heat and when mixed with oxygen, will be a louder, more turbulent (harsh) flame. Decreasing the acetylene flame will decrease the volume of heat and will be a quiet flame when oxygen is added.

Relationship to **distance**:

Do not use distance to control heat input. When welding, either hold the flame away from the piece or apply the flame to the work (⅛ inch from cone) and begin forming a molten puddle. The only time to be in between these two positions is when preheating the material before welding or when pulling the flame straight up from the end of a weld. When ending the weld, make sure to keep the molten puddle in the envelope for a moment to protect it while it solidifies.

Relationship to **angle**:

Your torch angle will determine where heat is placed. Incorrect torch angles can direct heat to places where it is not needed, or worse. It could be detrimental to a thinner section of metal or to only one part (piece) of a weld joint and not the other.

Relationship to **speed**:

Increased amounts of heat will require faster travel speeds. Adjust the amount of acetylene down for slower travel speeds and more control over the molten puddle.

TROUBLESHOOTING

Torch popping

When you are beginning to weld with oxy/acetylene, the torch will probably make a loud pop sending the molten puddle flying in all directions. This is a startling thing to have happen. You should understand why it happens and how to prevent it.

1. *Insufficient gas flow.* Make certain you are using the proper pressure settings for the tip used; a tip chart will tell you these settings. Be certain your acetylene cylinder valve is open at least ½ to ¾ turn and seated properly. Check for kinks in hoses or anything heavy laying on top of them. Add enough acetylene to the gas mixture; low amounts of acetylene can starve the torch.

2. *Dirty or worn out tip.* Obstructions in the tip will cause it to pop, along with soot, dirt, and slag on the outside. Even with proper care, tips eventually wear out. Replace the copper end of the tip every few years, or when the tip is no longer holding a proper-looking cone shape or flame.

3. *Overheating the tip.* Slow travel speeds and wide molten puddles reflect a lot of heat back onto the welding tip causing it to pop. Welding into the corners—like that of a tee joint—holding the cone too close to the work, and steep torch angles can all overheat the tip. Changing to a larger tip size will reduce the tendency of the tip to overheat, but at the same time will increase the overall volume of heat, which may not work well for the metal thickness you are welding. Use a proper welding technique to stop the torch tip from overheating before switching to a larger size.

Filler rod sticking to base metal

Again, you should know why this happens and how to prevent it.

- Use the correct-diameter filler rod for the metal thickness and tip size being used. Filler rods should be of equal or slightly less diameter than the thickness of the metal being welded.
- Make certain your molten puddle is both wide enough and well formed. Trying to add filler rod before the molten puddle is large enough will cause the rod to stick. Any rod that does melt off onto the plate will not be adequately fused to the base metal, and will mostly just be sitting on top of it.
- Change your work angle. Using a steeper torch angle will bounce more heat off the base metal and back into the end of the filler rod, preheating it. If your angle is too steep, too much heat will melt the end of your filler rod, putting a big ball on the end of it.

WELDING OUT OF POSITION

Vertical

If your progression of travel is upward (uphill), use a 45- to 55-degree travel angle; if your travel is downward (downhill), use an 85-degree travel angle. In either case, keep the tip of the torch pointed up toward the top of the plate. If necessary, vary the torch travel angle and heat input to prevent gravity from taking over. If the molten puddle becomes too big or too fluid, the molten metal can drip out and run down the work piece. If this should happen, reduce the amount of acetylene in the mixture or use a smaller tip.

Be certain to maintain a constant travel speed, and add filler rod frequently from above the weld pool. As a general rule, try to travel downward on thinner pieces (less than 1/16 inch) of sheet metal and upward on thicker pieces.

Overhead

This is the most difficult welding position only because you are less comfortable—your arms get tired quickly holding the torch and filler rod where they are needed. However, the distance, work angle, travel angle, travel speed, and amount of heat required are the same. The molten metal will not automatically drip out of the puddle. Liquid metal has cohesive (sticky) qualities that keep it in the weld pool as long as the molten puddle is not allowed to become too large. If the molten puddle becomes so big that it begins to run, move the torch slightly away from the puddle and allow it to cool before remelting it. Adding enough filler rod to the weld pool is also more critical, because the quench effect helps regulate the amount of heat in the molten metal.

ROUND TUBING OR PIPE

Get good at welding flat pieces of metal before attempting to weld round pieces. Work angles are easy to maintain, but your

Using a sturdy vise, clamp your weld coupon with the weld bead just above the jaws of the vise. Using a hammer, bend the coupon to a 90-degree angle. Remove your weld and use the jaws of the vise to bend the face of the weld 180 degrees. *Monte Swann*

travel angle must constantly be adjusted to keep your torch properly oriented. It is easy for your work angle to become too shallow, pushing the molten puddle ahead of where you want it, and bouncing much of the heat off the work piece.

TESTING YOUR WELDS

A good-looking weld with proper fusion and uniform appearance is usually strong, but not always. The only way to be certain is by testing. Any weld or braze on materials ⅛-inch thick or less can easily be tested for strength and soundness. A good weld or braze will be able to withstand a great amount of load without splitting or fracturing.

If your weld does fail, how and where it failed can give you a good idea of what problems may be occurring with your technique. Typically in welding mild steel, lack of fusion between the base and filler metal—either on one piece or both pieces of metal—will cause the weld to come apart.

REDUCTION

In the refining of metal ores, reduction refers to the fact that when iron ore is heated with a fuel like charcoal, which gives off carbon monoxide, it looses its oxygen molecules. The result is a separation of the pure metal from the impurities, which turn into slag and float to the top of the liquid metal.

In gas welding, the heat of the oxy/acetylene flame in contact with metal creates the same reduction phenomenon. Carbon dioxide is created, in this case shielding the weld pool. Some type of shielding atmosphere is required in all the manual welding processes covered in this book. All the other welding processes we will cover employ the use of a shielding gas and/or flux to protect the molten weld pool from being contaminated by the atmosphere's ever-present nitrogen, oxygen, and hydrogen gases.

If the face bend does not crack or break, use a pliers and hammer to flatten the weld. Use the vise and hammer to bend the root side of the weld back 180 degrees. *Monte Swann*

Notice on this root bend the seam is beginning to split on the left end. This is not unusual for a partial-penetration weld. If it were a full-penetration weld, this would not occur. *Monte Swann*

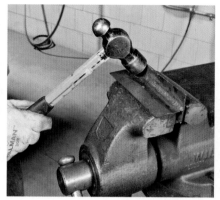

Use the hammer and vise to face-bend a tee joint, bending the bottom plate back 90 degrees. *Monte Swann*

Reinsert the weld in the vise as shown. Bend the top plate back 90 degrees. A weld with proper fusion will be able to withstand this type of punishment. *Monte Swann*

WELDING EXERCISES

I recommend completing these exercises in the order presented. The level of difficulty increases with each exercise, so become proficient with your new skills in one exercise before moving to the next. Wear the proper PPE. For the gas-welding exercises I wear a cotton/denim shirt for upper body protection and all-leather TIG welding gloves. Refer to Chapter 6 (Oxygen, Acetylene, and Other Compressed Gases) for details on start up and shut down procedures for using a torch, types of flames, and flame adjustment.

Before beginning these exercises you will need:

- Oxy/acetylene torch
- Proper-size welding tip (Victor No. 1 or 2)
- 14-gauge mild steel pieces cut approximately 2 × 6 inches
- Steel or stainless-steel wire brush to clean the metal before welding
- $\frac{1}{16}$-inch diameter 70S-2 mild steel filler rods
- Face shield or goggles with a No. 5 shade lens
- Pliers for handling hot metal
- Tip cleaner
- Quench bucket filled with water to cool and inspect your finished work faster
- Fire Bricks, which are densely packed, free of moisture, and designed not to blow up when in contact with a hot torch

EXERCISE 1

Pushing a puddle on a plate

The first important gas-welding skills to acquire are being able to see the molten puddle form and then to control it. It is much easier to take the time to master this skill now, before you need to add filler metal to the molten puddle.

Move the torch in circles; as you move along a path, this will translate into a series of circles. Keep the circles tight and small, no more than $\frac{1}{4}$ inch in diameter. If needed, draw two parallel lines $\frac{1}{4}$ inch apart on your plate with a sharp piece of soapstone; this will give you a path to follow.

When progressing forward, try not to skip ahead of your puddle. Keep the circles a consistent size and distance apart.

Each ripple in the example weld shows how far apart the overlapping circles are.

Last, manipulate the torch in a slow, even manner. Avoid manipulating the torch by stirring it; torch movement should be more deliberate than that.

Torch angles:

Work angle: 90 degrees (from side to side)

Travel angle: 45 degrees (in line with the weld axis)

Travel direction: forehand (same direction the torch is pointed)

After the molten puddle has formed on the end of your plate, concentrate on keeping your distance, angle, and speed consistent. Make sure you see the base metal actually melt and not just sweat. Use both hands when doing this exercise: one to hold the torch and the other to add stability by holding the wrist of your torch hand.

Begin this exercise by watching the molten puddle form below the tip of your cone. Stay in one place until you see the base metal turn into a liquid. *Monte Swann*

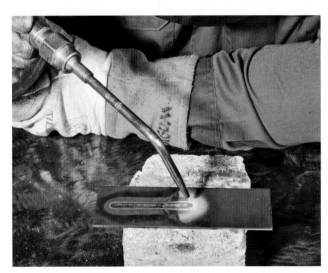

Once your puddle forms, begin circling your torch and slowly move forward. *Monte Swann*

This is a close-up of the molten puddle. Notice how it follows the inner cone of the flame. *Monte Swann*

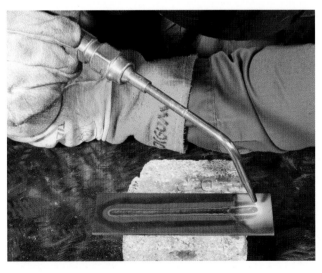

Be certain to maintain proper work and travel angles. Keep your travel speed consistent from beginning to end. *Monte Swann*

Stop short of the end of the plate. If you continue to push the molten metal off the edge of the plate, you will burn away the edge. *Monte Swann*

This is what your piece of metal should look like after this exercise; a weld bead of uniform width with fairly even ripples. There should be no melt-through on the backside of the metal. *Monte Swann*

GAS WELDING

Do not place your metal work piece directly onto another large piece of metal. The larger piece of metal will act as a heat sink, drawing heat away from the smaller work piece. This will create an unwanted quench effect and change the amount of heat required to create a molten puddle. Since welding is about regulating the temperature of the base metal, use fire bricks or metal pieces tacked into a couple of tee joints to isolate your work piece from any surfaces that will act like a heatsink.

When you are finished with the weld, turn the work piece over. There should be no evidence of burn-through on the back side of metal. If it looks like melting took place, or any areas have a convex appearance, your puddle was too hot.

EXERCISE 2

Full-open corner fusion weld with no filler rod (Autogenous weld)

Take some time when tacking the two pieces together. Before two pieces are welded together they need to be "tacked" together so they are held in alignment when heat from the torch is applied. Tacks are small welds, or small areas where the pieces are fused. Take some time to achieve good alignment with no gaps between the pieces. when tacking them together. Good fit-up is very important in this exercise.

Line up the edges of the metal, with the tips of the corners lined up, being careful not to overlap one edge on top of the other. You may need to flip and rotate pieces or straighten bowed pieces for the best fit-up. Since sheet metal stock is thin, bowed pieces can easily be straightened by bending. You can use two pairs of pliers, a hammer and vise, or place the piece of sheet metal between the work table and another piece and press down on the bowed side.

Torch angles:

Work angle: 90 degrees (from side to side of the weld axis)

Travel angle: 55 degrees (in line with the weld axis)

Travel direction: forehand (same direction the torch is pointed)

Using a circular torch motion, begin fusing the two pieces of metal together. Look closely at the molten pool and pay attention to the fusion taking place along the entire length of the joint.

Recognizing lack of fusion is key to this exercise. If your travel speed, torch motion, or torch angle is inconsistent, some areas of the weld will not come together and one or both edges will melt, but the two edges will not combine. This should be obvious while you are welding, so closely inspect the finished work piece for lack of fusion and learn to recognize it when you have a torch to metal pushing the molten puddle.

Carefully position two pieces of metal so each of the corners is in contact; try not to overlap them. Apply the flame as shown to fuse the two edges together. *Monte Swann*

It is only necessary to line up one end at a time. After the first side has been tacked, the coupons can be gently moved into alignment before tacking the other side. *Monte Swann*

If there is a gap between the two pieces, use a pair of pliers to hold them together while tacking. Be sure to hold onto the metal until the molten tack has solidified and cooled slightly. *Monte Swann*

As you begin to fuse the two pieces of metal together, make certain that complete fusion is taking place. *Monte Swann*

Use small circles and keep your torch at a 90-degree work angle. *Monte Swann*

Many people have a tendency to drop the travel angle as they weld. Be careful to maintain a proper travel angle throughout the entire length of the joint. *Monte Swann*

Keep your travel speed consistent. Moving too slow will cause the weld bead to sag and eventually it will burn through. *Monte Swann*

This is how your corner should appear after you are finished. Notice the edges have been smoothed into a radius and the weld has even ripples with no pin holes or lack of fusion. *Monte Swann*

EXERCISE 3

Bead on a plate with filler rod added

The goal of this exercise is to use the same welding skill learned in exercise 1 (Pushing a puddle on a plate), only now we will add filler rod to the weld bead. Filler metal builds up a weld, increasing its cross-section and adding strength to the joint.

The strength of a weld made with filler rod depends on proper fusion between the base metal and the filler metal. This is accomplished by a) adequately melting the base metal, b) adding the filler metal properly, and c) blending the filler metal into the base metal.

Torch angles:

Work angle: 90 degrees (from side to side of the weld axis)

Travel angle: 45 degrees (in line with the weld axis)

Travel direction: forehand (same direction the torch is pointed)

Filler rod angle: 30 degrees

Use the same circular torch movement, keeping distance and travel speed consistent. The angle of the torch will affect how the filler rod preheats. It is important to understand that by adding filler rod to the molten puddle, you are cooling the molten puddle. This quench effect of adding the filler rod is important in making a consistent weld bead. Adding filler rod, otherwise known as dipping the filler rod, on a frequent basis helps control the heat input into the base metal. Dipping the filler rod infrequently or inconsistently can cause the molten puddle to spread and areas of the base metal to overheat.

Begin this exercise the same way as pushing a puddle on a plate. Be ready to add filler rod once you begin circling your torch and the molten puddle becomes wide enough. *Monte Swann*

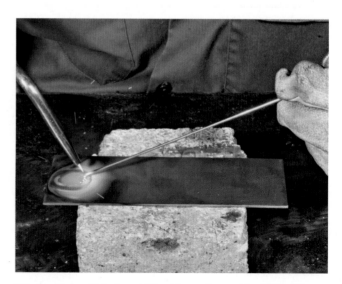

Keeping your filler rod close to the molten puddle will preheat the end of the rod, making it easier to add. Don't let the ball on the end become too big or drop onto the plate. *Monte Swann*

Dip your filler rod into the middle of the molten puddle. Adding filler material to the molten puddle has a quenching effect, cooling the temperature of the base metal and the weld pool. *Monte Swann*

GAS WELDING

After adding filler rod to the molten puddle, keep circling with your torch to blend the base metal into the filler metal. *Monte Swann*

As you circle the torch back, add filler rod to the middle of the molten puddle. As you circle the torch forward, blend the filler metal into the base metal with the heat of the flame. Try to dip the filler rod once every two or three circles. If you notice that no weld bead is building up, you may only be touching the molten puddle with the filler rod but not adding any. Don't be afraid to push the filler rod slightly into the molten puddle.

Important: Do not melt droplets from the rod into the molten puddle from above. This is improper welding technique and a bad habit to have. Droplets falling into the molten puddle can become oxidized from the atmosphere, contaminating the weld. In addition, this method of adding filler metal will not work for vertical and overhead joints.

These photos provide a better view of adding filler rod to the molten puddle and circling the torch forward, blending it into the base metal. *Monte Swann*

Learning to make consistent, small circles while dipping your filler rod often into the molten puddle is a skill that requires multi-tasking. If you are having trouble with this process, go back to the first exercise. Instead of using both hands on the torch, hold the torch in one hand and the filler rod in the other. Keep the filler rod close to the puddle you are pushing, but don't add any filler metal to the pool. *Monte Swann*

EXERCISE 4
Square butt joint in flat position
(Square groove weld)

Before tacking the two pieces of metal, you should decide how much of a gap to leave between the two pieces. In this exercise, I have tacked one end without a gap and the other end with a ¹⁄₁₆-inch gap. When starting out, it might be easier to orient your pieces as I did, but this will produce a weld that only partially penetrates the joint. A full-penetration weld is more difficult to make and requires leaving a gap the entire length of the joint.

Torch angles:

Work angle: 90 degrees (from side to side of the weld axis)

Travel angle: 45 degrees (in line with the weld axis)

Travel direction: forehand (same direction the torch is pointed)

Filler rod angle: 30 degrees

When tacking a square-groove butt joint together, put a gap of about ¹⁄₁₆ inch in at least one end of the work piece. If the two pieces are butted against each other over the entire length of the joint, contraction from the heat will pull your pieces out of alignment. Use a little filler rod when tacking the end with the gap. *Monte Swann*

Using the same technique as welding a bead on a plate, create a molten puddle and follow the seam of the joint. Keep moving the torch in small circles while adding filler rod to the molten puddle. *Monte Swann*

Be certain you are melting both edges of the joint. If the torch angle is incorrect, or if you wander away from the weld seam, there will not be enough heat applied to one side and the weld will be weak due to lack of fusion. *Monte Swann*

If needed, withdraw the torch from the end of the weld allowing it to cool slightly. Then, reapply heat at the end of the weld and add additional filler rod when it becomes molten for reinforcement. *Monte Swann*

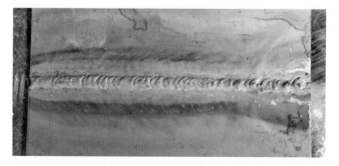

This is what your finished square-groove butt joint should look like. Although this weld may look good, it should be tested to prove that it is a sound weld. *Monte Swann*

EXERCISE 5
Lap joint in flat position (Fillet weld)

When tacking the two pieces together, try to overlap only one-third of each piece of metal so you have a large enough area of the lower piece to work with. A narrow area between the joint and the outer edge of the bottom piece will overheat quickly.

Remember, lap joints in the flat position are tipped at a 45 degree angle. Refer to the chapter on joint designs and welding positions for examples on how the joint, or workpiece, is to be oriented.

Torch angles:

Work angle: 85 degrees (from side to side of the weld axis)

Travel angle: 55 degrees (in line with the weld axis)

Travel direction: forehand (same direction the torch is pointed)

Filler rod angle: 45+ degrees (a different angle can be used to keep your gloved hand from overheating)

With lap joints, it is very important to use the correct torch angle. Keep in mind, the direction the torch is pointed is where the most heat is directed.

The exposed area of the lower plate is much thinner and smaller than the area covered by the upper plate. Begin by creating your molten puddle exclusively on the upper plate, then slightly change your torch angle to fuse the weld pool into the bottom plate.

Since the lap joint is a type of fillet weld, more heat is required in this exercise.

Tack the edges of a lap joint and use a piece of metal underneath to keep the top piece level. Use pliers to close any gaps between the top and bottom pieces when tacking the other side. *Monte Swann*

Begin this weld by creating your molten puddle on the base plate, just above the edge of the lower plate. *Monte Swann*

Begin adding filler rod to the molten puddle. Be certain the lower edge fuses into the weld. *Monte Swann*

Note how the puddle becomes C-shaped. Add your filler rod to the bottom of the C, which is at the root of the joint. This will help the top edge fuse into the bottom edge. *Monte Swann*

A lot of heat will be coming off the edge of the plate. Hold your filler rod so your hand is away from the heat. This is especially important when you are near the end of the weld. *Monte Swann*

83

In this photo, I have purposely changed the torch angle directing the heat to the bottom edge. Notice how this edge is melting away without a molten pool forming on the base plate. *Monte Swann*

It is possible to add filler rod at this time, however there will be a lack of fusion to the base plate. *Monte Swann*

Here is what the welded plate looks like. It is obvious where the torch angles were correct. The left end of the coupon is where I switched torch angles. You can clearly see that the weld position is different. The build-up on the bottom edge and lack of fusion to the upper plate is also apparent. *Monte Swann*

EXERCISE 6
Tee joint in flat position (Fillet weld)

The top plate of the tee joint should be perpendicular (90 degrees) to and in the center of the bottom plate.

The work piece is tipped 45 degrees from horizontal; tipping the plate in any direction will have a great effect on the placement on the molten pool. Welds in work pieces not positioned at 45 degrees tend to have unequal legs—more weld on one plate than the other.

Torch angles:

Work angle: 90 degrees (45 degrees from the top and bottom plates)

Travel angle: 55 degrees (in line with the weld axis)

Travel direction: forehand (same direction the torch is pointed)

Filler rod angle: 45+ degrees

While tacking the end of this tee joint, there was a small gap between the top and bottom plates. I was able to tack it by adding more heat, but melted the edge of the top plate creating a notch in the metal. *Monte Swann*

To hold the tee joint at a 45-degree angle, tack it to another piece of metal to support it and make it easy to reposition, if needed. *Monte Swann*

Begin this weld by applying the heat to both plates. If a molten puddle forms on one plate and not the other, angle your torch in the direction required to melt the other plate. *Monte Swann*

Begin adding filler metal to the molten pool. If there is not enough heat to add your filler metal, increase your torch angle to direct more heat to the molten pool. If there is still not enough heat, increase your amount of acetylene or change your welding tip to the next larger size. *Monte Swann*

This is a close-up of the weld pool and joint. To get proper fusion, be certain the molten puddle extends to the root of the joint, or the bottom seam. Adding too much filler metal may cause the weld to bridge the root of the joint. *Monte Swann*

Continue moving the puddle along, adding filler rod often. When you reach the end of the joint, be certain not to burn away the edge. *Monte Swann*

This is the completed weld. Notice that when welding a tee in the flat position, the joint acts like a trough keeping the filler metal from flowing out. *Monte Swann*

EXERCISE 7

Square butt joint in horizontal position (Square groove weld)

This procedure is more difficult because the welding no longer takes place on the tabletop. A horizontal groove weld is similar to welding out of position and is by far more difficult because of gravity's effects on the molten puddle. In addition, your body will be in a position that is less comfortable than sitting down with the work directly in front of you.

Although the pieces are vertically aligned, it is the seam of the joint that is the most important consideration. In the horizontal position, the seam is aligned horizontally. If the seam of the joint was vertically aligned, the weld would be in the vertical position. Refer to the chapter on joint designs and welding positions for more information.

Torch angles:

Work angle: 80 degrees (torch tip tilted up 10 to 15 degrees)

Travel angle: 55 degrees (in line with the weld axis)

Travel direction: forehand (same direction the torch is pointed)

Hold your torch in a vertical alignment to weld a square butt joint in the horizontal position. It is easier and more comfortable and the angle of the torch tip will naturally give you the correct work angle. *Monte Swann*

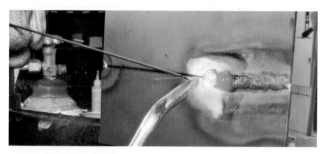

In this horizontal weld, you will have to compensate for gravity by using the correct torch angles and not overheating the molten pool. It is important to follow the seam of the joint, concentrating your molten pool in the middle. Notice here the plates are beginning to pull out of alignment. Putting a third tack in the center of the weld joint would have helped prevent this from happening. *Monte Swann*

Brace your arms as much as possible to steady the torch and filler rod. The molten puddle tends to follow the heat, but if too much heat is applied, it will sag and fall out of the joint. *Monte Swann*

Your torch angles should remain consistent, but the angle at which you hold the filler rod and add it to the molten pool can and should be changed so your glove doesn't overheat. Changing the filler-rod angle can also allow you to rest your arm. *Monte Swann*

On this completed weld, notice the slight undercut along the top edge of the bead. Undercutting the top plate is common in horizontal grove welds and should be avoided by using the correct torch angle, adding enough filler rod, aligning the plates correctly, and using the proper amount of heat. *Monte Swann*

CHAPTER 8
BRAZING, BRAZE WELDING, AND SOLDERING

When people think of brazing or soldering, they may think of a plumber joining pipes together, or an artist constructing a stained glass window. Many people will also equate welding with brazing or soldering. When I tell them that there is a big difference between welding and brazing, usually they are convinced until they hear about braze welding. Braze welding is not actually welding, nor is the brazing alloy, commonly called silver solder, actually used for soldering. It can become confusing for someone unfamiliar with how the different processes work. This chapter should help clarify the differences between welding, brazing, and soldering.

Industrial brazing is widely used to join metals in the manufacture of products such as bike frames and radiators. Large ovens or fixed torches are used and a conveyor belt moves the pieces along. As the parts pass by, heating elements melt strategically placed nuggets of filler metal (bits of filler alloy in the form of rings, washers, strips, and powder), joining large volumes of parts in an assembly line fashion. This chapter will concentrate on manual soldering and brazing with a handheld torch (torch brazing, or TB), discussing some of the common elements in each process, and explaining some of the uses and applications.

THE PROCESS

Brazing, soldering, and braze welding have a few things in common that set them apart from welding. The first and most important fact is, unlike welding, the base metal doesn't need to be heated to its melting point. Only the filler metal—which is usually an alloy containing copper and other metals and elements—melts. In welding, both the base metal and filler metal need to reach their melting points, become molten, and mix together in the weld bead (dilution) in order for fusion to take place. Brazing works by the filler metal adhering to the base metal. Brazing is more like gluing metals together instead of welding them. If done properly, brazed joints can be extremely strong.

GASES AND TORCHES

A variety of fuel gases can be used for brazing and soldering. Depending on how much heat is required, a fuel gas can be burned with air, compressed air, or high-purity oxygen.

Brazing and soldering require only a few tools. In the bottom left corner is a tubing cutter and wire brush used for cleaning the joint before applying the flux. Also pictured are various filler alloys and flux paste for brazing. *Monte Swann*

An air-acetylene torch (turbo torch) will provide a lower amount of heat than an oxy/acetylene torch, which may be an advantage when brazing or soldering small parts or very thin materials. MAPP gas, propane, and butane torches can also be used.

FLUX

In order for adhesion to take place between the base metal and the filler metal (filler alloy) a flux is usually used. The flux makes the filler alloy more fluid, allowing it to flow more easily into the joint. The flux also dissolves and removes unwanted oxides—which form quickly when metals are heated—creating a chemically clean surface. Although flux is a cleaning agent, it is not intended to remove contamination already present on the metal such as dust, dirt, grease, oil, and heavy oxides.

Fluxes come in powder and paste form, or baked onto the outside or wrapped inside of a pre-fluxed brazing rod. If you are using a paste flux, be certain to use the correct kind for the type of filler alloy you are using and for the type of joining process. For example, if by mistake you use a flux designed for soldering on a brazed joint, the flux will burn off long before the metal is hot enough to accept the brazing rod. A pre-fluxed rod takes the guesswork out of which flux to use and how much is needed. Find out from the manufacturer or a welding supplier which paste or powder flux is a match for your filler alloy. Some filler alloys, such as sil-phos, do not require flux.

PRECOATING (TINNING) AND WETTING

In soldering, brazing, and braze welding, the key to success is adhesion between the base and filler metals. If this bond (molecular union) does not take place, the joint is not successfully brazed or soldered.

Wetting refers to the liquid filler alloy, aided by the action of the flux, spreading out in a thin layer over the heated base metal. As soon as this wetting takes place, the filler alloy can flow on top of and/or into a joint. In some cases, the wetting action precoats the metal surfaces to be joined. Once the surfaces have been precoated, additional amounts of brazing material can be added to join and reinforce the joint. Non-wetting or de-wetting refers to when the solder or braze fails to spread out over the base metal and bond to it. In most cases this is due to the following:

- Improper type of flux is being used.
- Joint not cleaned well enough.
- Contamination deposited on the joint during or after cleaning.
- In braze welding non-wetting can also take place due to a lack of heat necessary to make the alloy flow.

CAPILLARY ACTION (ATTRACTION)

In soldering or brazing tubing and other shapes, the filler alloy will be drawn into a heated joint and will even flow uphill defying gravity. This is known as capillary action, also called capillary attraction. This physical force can be observed when two panes of glass spaced closely together are placed in a bucket of water. The water will be drawn up between the pieces of glass via capillary action. The same force is present when we use heat to melt and apply brazing or soldering alloys to a close-fitting joint. If the space of the joint is too tight, filler metals can't get in; if spaced too far apart, filler will adhere to only one side of the joint. In some brazing and soldering applications, it is critical that capillary action be complete throughout the spaces in between the tubing surfaces (faying surfaces) to create a leakproof seal. Sometimes this seal must withstand high pressures. In other applications, such as braze welding, capillary action may take place, but is not necessary or critical in creating a strong bond within the joint.

SOLDERING

Soldering is a low-temperature process, taking place below 840 degrees F. The equipment used for this process includes soldering coppers, electric irons, and torches. Most metals can be soldered together, including steel, galvanized sheet, tin, stainless steel, copper, brass, and bronze.

By their nature, the alloys used in soldering are soft and have a low melting point, therefore soldered joints have a low-tensile strength and are very sensitive to heat. Keep in mind solder is very fluid when molten, so spacing and fit-up is very important for good capillary action to take place. The space between the metals to be joined should be approximately 0.003 inch.

Tin/lead solders (copper-to-copper joint)

50/50 and 40/60 solders contain lead (50 percent and 60 percent lead mixed with tin). These are used for crafting and hobby applications, but should not be used with any lines in contact with drinking water. Be certain to wash your hands and clothing after handling these materials, as they are easily transferred to other surfaces and people.

Tin/antimony (soft) solders (copper-to-copper joint)

95/5 solder is commonly used in plumbing for water supply and drain lines because it doesn't contain lead. The solder has 95 percent tin and 5 percent antimony, a non-metallic element that acts as a cleansing and deoxidizing agent.

Silver bearing soft solder (copper-to-copper or copper-to-brass joints)

96 percent tin and 4 percent silver soft solder is free of lead and is often used to join suction lines and low-pressure components in heating, air conditioning, and refrigeration. Silver bearing soft solders have a greater strength than the other soft solders, yet still melts at a low temperature. This type of solder works well when fittings and tubing sizes are not perfectly matched, as silver solder is more elastic during soldering and will fill a gap between two parts more easily.

BRAZING

Brazing is a high-temperature process. Most brazing temperatures are between 1,200 and 1,400 degrees F (considerably hotter than soldering, but cooler than welding). Like soldering, brazing relies on capillary action to fill the joint with filler metal, so joints to be brazed are usually made with clearances of 0.001 to 0.010 inch. There are many different kinds of brazing alloys available to join almost any kind of metal or combinations of metals. The AWS organizes brazing filler metals into several classifications based upon the kinds of metallic elements contained in the filler metal. For example: BCu stands for *brazing copper* filler metal; BAlSi is *brazing aluminum-silicon*. Within each classification, different specifications are sometimes given, describing the exact content of each alloying metal. The percentage of alloying metals is indicated by a number or letter at the end of the classification, like the 4 in BAg-4. Which filler alloy you use will depend on the base metal(s) being brazed. The choice of flux depends upon the filler metal selected. Make certain to match your flux to your filler metal for best results.

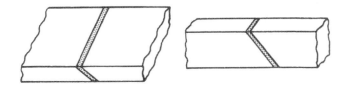

These are scarf joints used in brazing and soldering applications. Beveled surfaces are placed at parallel angles to increase the overall surface area of the joint. Scarf joints need to be carefully prepared to maintain the right-size gap along the entire joint, so proper capillary action can take place. For braze welding, scarf joints are a good option when all the reinforcement needs to be removed. *Welding Handbook Volume 2, Eighth Edition, Figure 12.13, Reproduced with permission from the American Welding Society (AWS), Miami, FL USA*

Silver braze (silver solder)

BAg-5 (B = Brazing, Ag = Silver, 5 = 45 percent silver, 30 percent copper, 25 percent zinc)

Minimum temperature at which brazing can take place: 1,370 degrees F.

This filler alloy is used to join copper, brass, steel, and stainless steel in any combination. Silver braze can also be used to join carbide cutting tips to steel tooling used in machining applications. The silver content in some silver braze alloys can be as high as 65 percent. A paste flux is required for proper wetting action to take place.

Sil-phos

BCuP-5 (B = Brazing, Cu = Copper, P=Phospherous, 5 = 80 percent copper, 15 percent silver, 5 percent phosphorous)

Minimum temperature at which brazing can take place: 1,475 degrees F.

A sil-phos brazing rod is used for copper to copper connections in a wide variety of applications. Because of the silver content in this filler alloy, *no flux* is required when brazing.

Joining metals with a brazing alloy has many advantages:

- The temperature required for brazing is less than welding.
- The base metal is not heated to a molten condition (liquefied).
- Less heat means the HAZ properties are not radically changed.
- Base metals tend to keep their mechanical properties, like ductility, malleability, and hardness.
- Less heat reduces and minimizes expansion and contraction forces within the base metal.
- Heat sensitive materials, such as medium- to high-carbon steels (carbon content of 0.35 percent or more) and hard, brittle metals (cast iron), are easily joined.

- Less preheating is required and post-weld heating or heat treatment is not necessary.
- Less heat input also means there is minimal distortion and warpage, especially on thinner pieces of sheet metal.
- A wide variety of dissimilar metals can be joined; one exception is that copper and copper alloys cannot be brazed directly to aluminum.
- MAPP gas, air-acetylene, and oxy/acetylene torches are very portable.

Disadvantages:

- Not used for joining or repairing parts exposed to high temperatures. Brazing filler alloys become weak at temperatures above 500 degrees F.
- Not used for joining steel parts under very high stresses.
- Filler alloy can be more susceptible to chemical corrosion.
- There will be a color difference between the filler metal and base metal.

BRAZING AND SOLDERING TECHNIQUE

The steps in soldering and brazing close fitting joints are similar. Follow the five-step plan below.

1. *Clean the joint area with sanding cloth and wire brushes.* Remove oxidation on copper tubing so it is bright and shiny. Be certain to clean both the male connection

This copper tube and copper fitting have been joined using a sil-phos brazing rod. Installing copper pipe and brass fittings are easy once you have practiced the proper technique. *Monte Swann*

Be certain to clean your work piece before soldering or brazing. Use a wire brush to remove any scale, rust, or dirt. Oil and grease can be removed with a solvent. Paint can be burned off with a torch and then brushed clean. *Monte Swann*

(outside of inner tube) and female connection (inside of outer tube). Clean all overlapping surfaces and a little beyond them, about ½ to ¾ inch away from where the parts come together. Any oil or grease can be removed with a solvent prior to brushing or sanding.

2. *Apply the correct type of flux immediately after cleaning the joint areas.* This reduces the chances that the metal will become contaminated by oxygen in the atmosphere, dust, and even the oils in your fingers if you touch the pieces again. If you are joining tubing, apply flux to the male connection only to minimize the amount of flux inside the tubing, which can become a problem later. On non-tubular connections, apply flux to the joining surfaces. Apply an even coat of flux to all the areas that have been cleaned, even ½ to ¾ inch away from the overlapping surfaces. Too little flux will burn away before the base metal is heated to a minimum temperature. Too much flux can clog and corrode the inside of tubular joints.

3. *Assemble the connection.*

4. *Heat the joint and apply the solder or brazing alloy.*

- Put a 90-degree bend in the filler rod about 1 inch from the end. This way you can tell how much filler alloy has been added to the joint. It is easy to add too much filler alloy to a joint. When soldering, the joint will appear to take lots of filler, when actually the excess is running out of the joint and onto the floor.

- Be decisive about adding heat. Hesitation and improper torch manipulation will burn out the flux making wetting the joint and capillary action impossible. There are two techniques for applying heat:

1. Apply heat to the joint so the inner cone of the flame is just above, or just touches the base metal.
2. Hold the flame 2 to 3 inner cone lengths away from the joint. Using this technique on tubing allows the outer envelope to wrap around the tube.

No matter which method is used, managing the heat input to the base metal is the most important consideration. Once heat is applied to a joint, the heat should not be completely removed until finished.

- Keep the torch in constant motion, sweeping to both sides of a joint, concentrating heat on where the joint overlaps. Move to the back, left, and right sides of tubing from where you started, applying the heat all the way around the tubing.

- Thicker pieces, such as brass fittings, require more heat. Keep this in mind and be careful not to overheat thinner sections.

- Begin testing the joint by applying the filler rod at the point where the overlapping pieces meet. Try not to pull the torch entirely away from the joint; instead, pull it back and a little to the side.

- When the proper temperature is reached, the filler metal will flow into the joint by capillary action. If the filler rod doesn't flow into the joint, pull it away and reapply the flame.

- Once the filler alloy begins to flow, add just enough to fill the joint.

3. *Clean the joint.* Some fluxes used in brazing and soldering are corrosive and can cause oxidation. When finished soldering, clean the joint with soap and water. When finished brazing, chip the flux residue from the joint. You need to remove this hard, glass-like substance, as it may cover a weak spot in the braze, which can eventually leak.

Tip: What to look for when soldering

Soldering is a low-temperature process. You can tell copper tubing is hot enough when different colors begin showing on the surface where the heat is applied. These temper colors are blue-green-straw. If you see the tube itself begin to turn red or orange, too much heat has been applied for soldering. Also, look for the paste flux to liquefy.

Tip: What to look for when brazing

Brazing will require more heat. Copper tubing will be cherry red in color at the proper temperature for brazing. Look for the flux to turn clear.

Tip: When joining at a waterline, sometimes it's difficult to keep the tubing dry enough to braze. Stuff the inside of each tube with soft white bread. That way it will absorb the water, allowing enough time to join the metals, before dissolving away. Also, hardware stores sell absorbent beads that can be stuffed into the lines. The beads look like bath balls, and come in different diameters for each tubing size.

MAKING REPAIRS

If you find you have a leak in the line you just brazed, it is possible to reheat any brazed or soldered joint with a torch and melt out the filler alloy. When it comes apart, reclean the joint, reapply flux, and make another attempt at brazing the joint together. Copper becomes very soft and loses some mechanical properties when reheated multiple times, so this solution will not always work. It may be necessary to cut out a leaky joint and use fittings to make a repair.

ORDER OF OPERATIONS

On parts where more than one joining process will be used, take a minute to think about the amount of heat required for soldering, brazing, and welding. If you are brazing and soldering on adjacent pieces of copper tubing, do the higher-temperature brazing work first, then do the lower-temperature solder work. A solder joint can be melted by the heat of brazing, and a brazed joint can be melted by the temperatures required for welding.

BRAZE WELDING

Braze welding is a great way to join and repair all types of cast-iron and steel parts, but with less heat input than is required for welding. This process can also be used on copper and nickel alloys, on pot metals, and to join combinations of dissimilar metals. Braze welding can be a simple way to join large joint areas or build up worn surfaces, such as a gear tooth. Since braze welding does not rely on capillary action in order to work, the amount of clearance between the parts is not critical. What is important is that there is good adhesion between the filler alloy and base metals, and that adhesion takes place over a large enough surface area for the strength requirements of the joint.

Braze-welding filler alloy

Low-Fuming Bronze

RBCuZn-C (RB = Rod braze welding, Cu = Copper, Zn = Zinc, C = 60 percent copper, 38 percent zinc, 1 percent tin, 1 percent iron)

Minimum temperature at which brazing can take place: 1,630 degrees F

Minimum tensile strength: 50,000 psi

BRAZE-WELDING TECHNIQUE

1. Prepare the base metal for braze welding by using a wire brush to clean off any contaminants, such as rust and dirt.

    - Remove mill scale and casting skin (coating on the outside of cast-iron pieces) in the joint, and ½ to ¾ inch away from the outside edge of where the filler alloy is to flow.
    - Oils, paints, and other unwanted materials can be burned off by sweeping the area with an oxidizing flame. Protect yourself from any harmful or toxic fumes.
    - Thinner pieces—⅛-inch thick or less—can be braze welded without beveling.
    - On thicker pieces, prepare the joint by grinding a bevel in both pieces at 45 to 60 degrees. This will produce a groove angle of 90 to 120 degrees, which is much wider than what is usually required for fusion welding. This wider angle is used to increase the surface area on which adhesion can take place.

2. Use fixtures or clamps to align the pieces and keep them in place during heating and cooling.

3. Preheat the base metal to about 200 degrees F. This temperature is described as a black heat, which is less heat than the amount needed to turn the metal a red color. Large pieces especially need to be preheated to minimize the quench effect of the metal adjacent to the joint. Run the torch briefly over thinner pieces to bring the metal up to room temperature (70 degrees) and burn off any moisture on the surface.

4. Precoat the surfaces (tinning)

    - Apply the proper amount of heat to the base metal. When it just begins to glow a dull red color, begin applying the filler rod. Don't overheat the base metal; begin testing the heat with the filler rod before the base metal gets too hot.
    - Apply small amounts of filler rod and let it spread over the joint. If the joint is overheated or the bronze filler metal gets too hot, it will boil and burn, giving off a white smoke. If this happens, let the base metal cool down before continuing. Coat the entire surface of the joint. Remember, braze welds are strong only because of proper adhesion. Capillary action is not necessary in braze welding.
    - If you see areas in the pool where the bronze is flowing around an area and not adhering to that spot, the base metal is not hot enough.
    - If you think an area has not been heated enough, move the torch back over the spot. Even when reheating a finished bead, if there were voids left in the filler metal, they will open back up.

Keep heating and circling the torch in the area until the bronze flows back into the cold spots and completely adheres.

5. Apply subsequent layers if necessary, filling up joints and building up surfaces. Additional braze-welded layers will add a great deal of reinforcement and strength to the joint. Carefully observe that good fusion is taking place between the new bead and precoating or previous layer. Always add the prefluxed filler rod to the braze-weld pool. Never add filler material to the joint by melting droplets off the end of the rod and letting them fall into the pool.

6. When possible, position your work at a slight angle and travel uphill. This will help the bronze adhere to the base metal and prevent the liquid bronze from flowing ahead of the heated area.

7. If a work piece can be brazed from both sides, create a V-groove (90–120-degree groove angle) on one side only, to a depth of just over half the metal thickness. The other sides of the pieces are left in their original condition, so it is easy to fit the pieces back together. For example, the parts of a broken casting may need to line up in a certain way. Preheat, precoat, and braze weld the first side. Then turn the piece over and grind out a V-groove on the other side. Grind down to the root of the previous braze weld; the yellow-gold color of the filler metal will be easy to see. Proceed to precoat and braze weld the second side. If done correctly, this joint will be very strong.

8. When repairing broken cast iron, sand the V-groove along the cracked surfaces. Castings should be slow-cooled by burying them in sand or kitty litter or wrapping them in a fire blanket. Castings are easily machined after being rejoined by braze welding.

FILLET BRAZE WELDS (LAP AND TEE JOINTS)

All types of joints can be successfully braze welded. Practice tee and lap joints in the flat position first. Remember, these joints require more heat to braze weld successfully, so a little preheating may help.

On lap joints in the flat position, apply more heat to the top plate and avoid overheating the bottom plate. Allow some bronze material to flow up over the edge for additional reinforcement. On tee joints, try to distribute the filler metal evenly on both the top and bottom plates.

To test these types of joints, clamp them in a vise and hammer the metal pieces toward the brazed side to expose the root of the joint. If they break apart, look at the root to see if there are voids or areas that lack adhesion.

It is possible to root bend, or back bend, a braze-welded, mild-steel tee or lap joint and not have it break, in which case everything was done correctly. Sometimes joints will split in places and stay together in others, indicating part of the joint had proper adhesion and enough reinforcement to hold together.

HORIZONTAL AND VERTICAL JOINTS

It is possible to braze weld both horizontal and vertical joints, although they are more difficult since the bronze filler can easily run out of the joint if it gets too hot. Controlling the amount of preheat and total heat input is more critical when gravity becomes a factor. On horizontal joints, precoating the joint can reduce the heat required for final brazing.

Use multiple passes to obtain an even amount of bronze filler on each part of the joint. On vertical joints, build a small shelf of filler metal at the bottom of the joint and travel upward, building up layers as you go—like climbing each step of a ladder. Test joints for strength and soundness.

BRAZE-WELDING EXERCISES

I recommend completing these exercises in the order presented. The level of difficulty increases with each exercise, so become proficient with your new skills in one exercise before moving to the next. Wear the proper PPE; I'm wearing the same gear I had on for oxy/acetylene welding.

You will need:

- Oxy/acetylene torch with a Victor No. 0- or 00-size tip (other torches can be used instead)
- Braze welding rod, LFB–FC $\frac{3}{32}$ × 36 inches (low-fuming bronze-flux coated $\frac{3}{32}$-inch diameter and 36-inch length - AWS designation RBCuZn-C)
- 14-gauge hot- or cold-rolled mild-steel pieces cut approximately 2 × 6 inches
- Face shield or goggles with a No. 5 shade lens
- Pliers, tip cleaner, and wire brushes

Before you begin, be certain your equipment is functioning properly. Use a smaller-sized tip, one or two sizes smaller than the tip used for gas welding the same metal thickness. Control heat input by using proper gas pressure and flame settings, and follow all precleaning procedures and braze- welding techniques. Use the proper filler alloy and select the correct type of flux, if necessary. If the set-up is incorrect, it will be impossible to make a good braze or solder joint.

EXERCISE 1
Bead on a plate

If you are used to the amount of heat needed for gas welding, it may take some time to adjust to joining metals with less heat. This exercise will help you recognize the amount of heat required for braze welding and allow you to practice controlling the amount of heat input.

Add filler rod to the piece by quickly moving the rod back and forth into the heated area. If done properly, the braze weld will make a soft sizzling sound as it spreads out and adheres to the base metal. No precoating (tinning) is required for braze welding thinner materials $\frac{1}{8}$ inch thick or less. The same principles of distance, angle, speed, and heat that apply to gas welding also apply to braze welding.

Apply the oxy/acetylene flame, keeping the cone approximately ¼ inch away. Heat the steel to a dull red and avoid liquefying the metal. *Monte Swann*

Begin circling the torch and add small amounts of brazing rod. Observe the brazing rod as it flows onto the base metal. In this example, the cone is too close to the filler rod and base metal. *Monte Swann*

Continue adding filler metal and push the bead along the plate. Notice the flux coating melting off the end of the rod and flowing around the bead. It is normal for this to happen. Try to look past it, and look instead at the bronze filler adhering to the base metal. *Monte Swann*

When you approach the end of the plate, it may overheat. If this happens, pull the torch away and allow the piece to cool before reheating and adding more braze to the last few inches. *Monte Swann*

When the zinc content of the bronze rod burns off, it creates zinc oxide, which shows up as a white residue on either side of the brazed joint. Notice parts of the bead have a glassy appearance; this is a byproduct of the flux coating. This residue should be chipped and brushed off. *Monte Swann*

Distance

Keep your heating source ¼ to ½ inch from the cone to the work. Use the envelope, instead of the cone, to heat the base metal with a neutral or slightly oxidizing flame.

Angle

Forehand: 45–55 degrees (traveling in the direction the torch is pointed)

Backhand: 85 degrees (traveling in the opposite direction the torch is pointed)

Work angle: 90 degrees to base metal

Speed

Travel as fast as the bronze filler material will flow with good adhesion to the base metal. Don't stay in one place too long or travel too slowly; the amount of heat input is more critical with this process.

Heat

Use smaller-sized tips, like a Victor No. 0 or 00. Use less acetylene in the mixture to produce a soft flame, which is good for brazing. Harsh flames should be avoided.

This is a cold bead. The base metal has not been brought to the correct temperature in order to make the bronze flow. A bead without a smooth transition to the base metal indicates lack of adhesion.
Monte Swann

The bead should be of uniform width and height, with smooth, fine ripples and a bright bronze color. The edges should flow into the plate with no signs of a lack of adhesion.

There are a couple of clues that indicate overheating:

1. *Pits or porosity in the braze bead.* Look for a small line of pores down the center.
2. *Zinc oxide residue*—a white, powdery substance created when the zinc in the filler metal burns off. May be on either side of the bead, but should *not* cover the bead.

3. *Cracks in the base metal after cooling.* These cracks occur when the base metal becomes semi-molten and bronze filler metal works its way in. Because the bronze has a lower melting point than the surrounding base metal, it stays liquid while the base metal begins to solidify. The base metal is then forced apart by the liquid bronze, creating a crack, usually down the center on the opposite side of the bead or on the same side directly next to the bead.

EXERCISE 2
Square butt joint in flat position
If you can't use clamps or a jig, just tack the two pieces together. Either way, leave a root opening (gap) of about ³⁄₃₂ inch the entire length of the joint. Too large a root opening will cause the filler metal to flow through the joint; too small and the brazing material will not flow through at all.

Use the same DASH principles as in exercise #1. Maintaining a 90-degree work angle is more critical with a butt joint. Tilting your torch more toward one plate than the other will put more heat to one side. The result is a lack of adhesion on the side the torch was pointed away from.

After the first pass, if the bead profile is almost flush and the braze material has flowed through the joint to the backside, make another pass over the previous bead to add reinforcement. If, during braze welding, the bronze filler keeps going into the joint no matter how much you put in, back off the heat and check the backside. The filler may be flowing through the joint and building up there.

Test your braze weld for strength. It should hold together, even after several face and root bends are made. Look for lack of adhesion and reinforcement in joints, which split or come apart.

The heat of the torch in the brazing process can move work pieces out of alignment. Any pieces to be braze welded need to be clamped or tacked together (as shown) to maintain any spacing and fit-up requirements.
Monte Swann

This square butt joint has a 3/32-inch gap along the entire length. At the beginning of this braze, not enough filler alloy was deposited. This could cause a weakness in the joint due to lack of reinforcement. *Monte Swann*

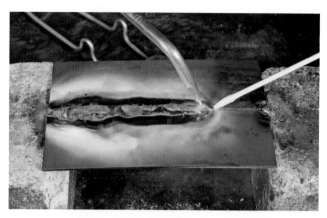

Circle the torch and add braze as required. Observe the brazing filler rod sucking into the joint. Also notice the wide area where more heat was put to the plate and more brazing rod deposited. Try to keep your bead width as consistent as possible. *Monte Swann*

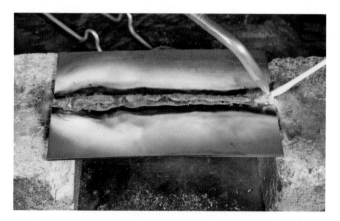

You always want to add enough reinforcement at the end of a brazed joint for strength. Don't hesitate to pull the torch back and let the metal cool if it overheats. If you finish without enough reinforcement, a split or crack is likely to occur. *Monte Swann*

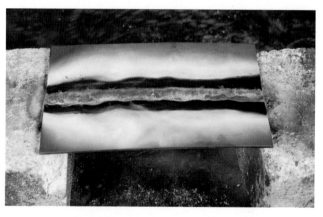

The edges of your braze should have a smooth transition into the base metal, which indicates proper adhesion. If the bead height or height of reinforcement is not adequate, another pass can be made over the top of the first bead to increase the strength of the joint. *Monte Swann*

This is the backside of the brazed joint. Notice the bronze filler alloy has been drawn through the gap and has flowed around the backside of the joint. Ideally, you want a more consistent flow the entire length of the joint. Notice that the top piece of metal was previously used in the gas-welding chapter for exercise 1. Reuse your metal as much as possible while you are learning how to weld. *Monte Swann*

You can test your brazed joints in the same way as your welds with a hammer and vise. Knock the face of the braze weld over 90 degrees and look for any lack of adhesion between the bronze and base metal. *Monte Swann*

Hammer the face of the coupon back to flat and reinsert the joint into the vise. This time, hammer the root of the braze weld over 90 degrees. *Monte Swann*

Use the jaws of the vise to continue your root bend to 180 degrees. A properly executed braze weld will be able to withstand this amount of load without coming apart. Notice the bronze filler runs the entire length of the joint. *Monte Swann*

CHAPTER 9
INTRODUCTION TO ARC WELDING

About 200 years ago, people discovered a controllable electric arc. The arc is like a little battery-powered bolt of lightning sparking between two solid carbon rods. At the time it was nothing more than a parlor trick. Fifty years later, electric generators were developed. The bright light produced by an arc was used to light the streets, replacing older whale-oil lamps.

Carbon-arc welding first came in 1880. Several patents for various arc-welding processes were awarded during that time for both carbon-arc and metal-arc welding. Carbon-arc welding used solid carbon rods and the heat of an electrical circuit to fuse two pieces of metal together. This type of welding was limited by the fact that the process introduces small particles of carbon into the weld metal, making the joint hard and brittle.

Metal-arc welding had a greater potential for development and would later become shielded metal arc welding (SMAW). The difference between it and carbon-arc welding was that a bare metal rod now became part of the electrical circuit and carried the heat required to create a molten puddle in the base metal. At the same time, the metal rod melted and became part of the weld. The problem with early metal-arc processes was that welds were porous, brittle, and weak. For a while

it was a mystery why rusty metal rods were found to work better than new, clean ones. Only later was it discovered that an iron oxide coating (rust) provided limited shielding effects from the atmosphere.

During the first decade of the twentieth century, two companies from Germany brought the metal-arc process to the United States and inadvertently made the process available to the general public. In a legal battle over patent protection, one company sued the other. In the court's final decision, both companies lost all patent rights to the metal-arc welding process. From 1900 to 1928, several welding manufacturers were established and began experimenting and perfecting the coatings on the outside of the metal-arc electrodes. By 1929, the Lincoln Electric Company mass-produced the modern shielded metal electrodes and made them available to the general public. By 1930, once the flux was perfected, there were oceangoing ships being built exclusively by welding.

Although in development during the same period as SMAW, it wasn't until 1941 that gas tungsten arc welding (GTAW) was used on a mass scale in manufacturing. During World War II, GTAW was used to join aluminum, magnesium, and stainless steels in the production of aircraft. Like carbon-arc welding, GTAW uses an electrode that is not consumed in the welding process and does not become part of the weld. The electrode for GTAW is made of tungsten instead of carbon. Unlike carbon-arc welding, the tungsten electrode does not deposit carbon to the base metal. GTAW welds are strong and ductile because a shielding gas is used to protect the weld from atmospheric contamination. The first shielding gas used was helium, so GTAW began its life being called heliarc welding.

In 1948, a new process was developed—a cross between GTAW and SMAW. This new process involved a metal electrode that was consumed by the welding process, becoming part of the weld, and a shielding gas to protect the molten weld pool. In addition, instead of the electrode being a stick or rod, it was a spool of wire continuously fed by a motor and drive wheels. The new process was called wirefeed welding, also known as gas metal arc welding (GMAW). The shielding gas used at the time was argon, which was expensive and limited wirefeed applications, until a cheaper substitute came along in 1953: carbon dioxide. Many wirefeed processes were developed in the 1950s, including short-circuit transfer with small-diameter wires (smaller than 0.045 inch in

A welding program at your local technical college will have a large number and wide variety of machines for you to try out. Most schools will have classes open to the public giving you access to machines you may not normally have. *Monte Swann*

diameter) known as micro-wires, and wires with flux cores (FCAW-G and FCAW-S). Wires with a powdered metal core were introduced as late as 1976.

Today SMAW, GTAW, GMAW, and FCAW are widely used processes in manufacturing plants, construction sites, and metal fabrication shops. Individual sections of this book are devoted to each of these processes, explaining their advantages and disadvantages.

ARC WELDING DEFINED

Since SMAW was the first modern arc-welding process to be developed, it has inherited the name arc welding. If someone uses the term arc welding they usually mean stick welding or SMAW. In addition to SMAW, there are many other arc-welding processes. GTAW and GMAW are also arc-welding processes. They all have one thing in common: the use of electricity to generate heat and create a molten puddle.

Resistance

There is a small, fine wire in traditional light bulbs. As electricity flows through the bulb, it hits that small wire, which becomes so hot it begins to glow. The reason the wire gets hot is because a large amount of electricity is trying to flow through what is suddenly a very small channel. This creates abnormal resistance in the circuit. All circuits have some resistance; it is measured in ohms (Ω). Resistance creates heat and the heat creates light.

Welding also relies on resistance in the circuit, except instead of suddenly being made to go through a small wire, the electricity crosses a gap. The gap is kept small enough that electricity can jump across it and the circuit remains uninterrupted. As electricity crosses the gap, it creates an electric arc, making arc welding possible. The electric arc used in welding has been measured at 10,000 degrees F—nearly as hot as the surface of the sun.

Electrodes

In welding, an electric arc is created between the base metal and the electrode. The base metal is connected to a work clamp (ground clamp), and the electrode is connected to an electrode holder. Both are attached to cables called leads. The electrode lead and work clamp lead both connect back to the welding machine, so the base metal and electrode each become part of the electrical circuit used in arc welding. Electrodes fit into one of two categories: consumable or non-consumable.

In SMAW, the electrode (also known as the welding rod or stick) is consumed during the welding process. This means that the filler metal is part of the electrode itself. This filler metal is super-heated and liquefied at the end of the electrode. The liquid metal crosses the arc gap inside the arc stream and is deposited into the base metal, which is also liquefied by the heat of the electric arc.

In GMAW, the electrode is in the form of a solid wire wound on a spool. This wire is continuously fed by the machine and is consumed by the welding process, eventually becoming most of the finished weld bead. Although electrodes can contain filler metal, they are not called filler rods.

Non-consumable electrodes are different because they never are consumed in the welding process and do not become part of the weld (unless by mistake). Non-consumable electrodes still carry the heat of the electric arc, but are made from materials that can withstand a great amount of heat. In GTAW, the metal tungsten is used as the electrode. Tungsten has a very high melting point and will not melt before the base metal does. This allows us to create a molten puddle on the base metal without adding any metal from the electrode. Additional metal is added using a filler rod. Although GTAW filler rods and electrodes may appear to be similar, they are different, because filler rods never become part of the electrical circuit.

ELECTRICAL CIRCUITS, OR PATHS TAKEN BY ELECTRICAL CURRENT

When the electricity flows out from the power source, through the cables, and returns back to the power source, this path is known as an electrical circuit. An interruption in this circuit will not allow the electricity to flow all the way back to the power source.

The best example of this is a light switch. When the switch is off, it creates an opening in the circuit and the flow of electricity is interrupted. Since the flow of current (electrons) never reaches the light bulb, it stays unlit. When the switch is turned on, the opening is bridged, allowing electricity to flow through to the light bulb and then back to the switch. The light bulb uses a certain amount of energy or electricity to create light, so there is a drop in the amount of energy leaving the light bulb. As long as the switch is closed (in the on position) and electricity is supplied to the circuit, the light will stay lit. Welding works the same way; the circuit must be complete in order for the welding arc to stay lit.

Voltage: The force that causes electrical current to flow in a circuit.

Amperage: The amount, or rate, of electrical current flowing in a circuit.

Think of electricity flowing through cables as being like water flowing through a hose. At one end of the hose is a pump, which pushes water through the hose. The size and capacity of the water pump is like voltage. The volume of water flowing through the hose is like amperage. If the size of the pump is matched to the diameter of the hose (proper voltage for the amount of amperage), the flow of water will be consistent and come out the other end at an efficient rate. If a small pump is used for a large-diameter hose (voltage too low for the amount of amperage), water will only trickle out the other end. If a large pump is used for a small-diameter hose (voltage too high for the amount of amperage), then water will be sent forcefully through the hose, but the amount of water coming out the other end will never be more than what the diameter of the hose will allow.

There is a relationship between amps and volts that needs to be maintained in order for the arc-welding process to be efficient and work effectively. If you kink the garden hose, water pressure builds up at that point, possibly bursting the hose or breaking the pump. The kink is like unwanted resistance in an electrical circuit that causes a backup in the flow of electrons and a buildup of heat.

AC/DC

In arc welding, electricity is generated by a power supply and sent through cables called welding leads. The flow of electricity (electrons) is always from negative to positive. With direct current (DC), the positive and negative poles maintain their positions and the current flows in one direction. Alternating current (AC) is continually and rapidly reversing because the poles alternate locations. The charge at the power source is what causes the poles to switch locations.

Alternating current (AC)

AC alternates between positive and negative poles. When electricity is sent through power lines and comes to an outlet in a building, it is in the form of AC. As its name implies, this type of current (amperage) switches directions between the two poles of positive and negative electrical charge every 1/100 or 1/120 of a second. Therefore, one full AC cycle takes place either 50 or 60 times a second, which is a frequency of 50 or 60 Hertz (Hz). Some electrical appliances, like clocks and electric motors, made for use in North America (where the standard frequency is 60 Hz) won't work in Europe unless an adaptor is used to change the frequency. If a machine or appliance is rated for both 50 and 60 Hz, it will work anywhere in the world, as long as the correct voltage and plug connection is used.

When running AC, it will not matter how the leads are attached to the machine because the electricity is constantly switching directions.
Monte Swann

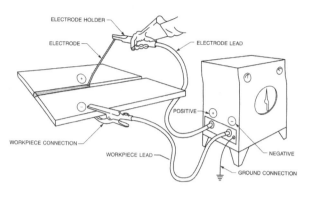
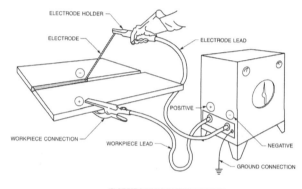

(A) DIRECT CURRENT ELECTRODE POSITIVE

(B) DIRECT CURRENT ELECTRODE NEGATIVE

AWS A3.0:2001, Figure 34, Reproduced with permission from the American Welding Society (AWS), Miami, FL USA

In arc welding, AC output current is used in a variety of applications. In GTAW, AC current is used when welding aluminum and magnesium. In SMAW, certain electrodes can be used with AC current at a lower amperage than required for the same rod using DC. AC also eliminates the problems associated with arc blow. AC current is never used in GMAW. Also, AC can be transformed into DC.

Direct Current (DC)

DC flows in one direction only. It is assumed that electricity flows from the negative pole to the positive pole. The way the welding leads are hooked up to the welding machine will determine which way the current will travel through the electrode holder and work clamp.

Polarity

Polarity indicates the direction of flow in a circuit. There are two types of polarity.

1. *Reverse Polarity*
   Also known as:

   * Direct current reverse polarity (DCRP)
   * DC+
   * Direct current electrode positive (DCEP)

If the cable attached to the electrode (holder) is connected to the positive terminal on the machine and the work clamp is attached to negative, the electricity will flow through the base metal first, jump across the arc gap to the electrode, continue up the electrode, and back to the machine. A majority of the heat is concentrated in the part the arc is jumping to. In reverse polarity, a majority of the heat is concentrated in the electrode.

2. *Straight polarity*
   Also known as:

   - Direct current straight polarity (DCSP)
   - DC–
   - Direct current electrode negative (DCEN)

If the cable attached to the work clamp is connected to the positive terminal and the electrode holder is attached to negative, the electricity will flow straight down the electrode first, jump across the arc gap to the base metal, and back to the work clamp and machine. In straight polarity, a majority of the heat is concentrated in the base metal.

DC is commonly used in welding. Either DCEP or DCEN are used in SMAW, depending on the type of electrode and metal thickness being joined. Only DCEP is used in GMAW. In GTAW, usually only DCEN is used to weld all types of metals, except aluminum and magnesium.

ARC WELDING SAFETY
Arc light

Think of the electric arc as a small sun, producing a lot of heat and emitting a blinding light. Arc light contains ultraviolet (UV), infrared (IR), and visible light rays, making it very intense. To actually see the weld take place, a shaded lens is required. These lenses are too dark to see through without the

If the work clamp is connected to the positive terminal, and the electrode holder to the negative terminal, the polarity is DCSP.
Monte Swann

light of the arc. That's why some people say welders do their work in the dark.

There are specific hazards associated with arc light. One of them is not blindness. The idea that arc light will make you go permanently blind is a common misconception, along with the notion that you cannot wear contacts while welding because they will melt to your eyes. Arc light is hazardous, but not for either of these reasons. The real danger from exposure to arc light is a condition known as arc flash.

Think of arc flash as a sunburn on the surface of your eye ball. Like your skin, your eyes need to be exposed to the light for a certain amount of time. If you step outside for only a moment and then go back indoors, you won't get burned. Stay outside all day without protection, though, and you'll burn for certain. When you do go outside, the sun can burn you regardless of whether you are looking directly at it or not. The same is true of arc flash—you need not be looking directly at the arc light in order to burn your eyes.

Fortunately, arc flash is a non-permanent condition and will not cause blindness. Arc flash is very uncomfortable, but will not hurt right away. Instead, sometime later in the evening you will feel as if someone is pouring sand directly in your eyes. There is nothing to cure you; only time will heal the surface of your eyes. It is best to avoid arc flash.

1. Prolonged exposure to arc light can result in arc flash. Pick the proper lens shade for the type of work you are doing. Avoid looking at the arc or flash unless equipped with the appropriate lens. The proper shade will depend upon the type of arc welding and the size of the welding electrode being used. A minimum shade of No. 9 should be used for arc welding. If you see spots after welding, your shade was not dark enough. In this case go to the next shade up, for example, from a No. 10 shade to 11. See the chapter on safety for more information about shaded lenses and welding helmets.

2. Be sure your helmet and goggles are in good condition. Cracks or holes permitting penetration of arc light can lead to arc flash. Remember: you don't need to be looking directly at the light to get burned.

3. Caution others about watching the arc light without proper eye protection. People are attracted to the bright light, especially children and pets. Protect bystanders from being burned by using light-blocking curtains, making them wear a welding helmet, or keeping them away from the area.

4. Over time, exposure to IR and UV light can cause cataracts to develop.

5. Wear protective clothing to protect skin from intense rays; any exposed skin can be burned by IR and UV light. Avoid wearing any white clothing while arc welding. White tee shirts for example will reflect the arc light back up onto your face.

INTRODUCTION TO ARC WELDING

Body protection

1. Always wear all-leather welding gloves that can be thrown off easily. This is in case your gloves accumulate heat from welding, or a hot spark, piece of hot metal, or hot slag accidentally goes down you glove.

2. With GTAW, there is less heat in the surrounding area and virtually no sparks, so some welders will wear only one glove or no gloves at all. Their hands will smell funny after welding because the arc light has been burning their bare skin. Very thin TIG gloves can give you all the manual dexterity you need for welding without exposing your skin to the dangerous light.

3. Some arc-welding processes, such as SMAW and FCAW, are more aggressive in generating sparks, hot slag, and molten metal. Where necessary, wear leather aprons and arm coverings to protect yourself. This is especially important when welding out of position (vertical and overhead).

4. Wear high-top shoes rather than undercut shoes. They should be at least ankle height, again because of the hot metal and slag.

5. If a helper is used, he/she should be protected the same as the welder. Have extra gear around for people helping you out.

6. Always wear safety glasses at all times, even under your helmet. Never chip slag without safety glasses on. The slag is very hot right after welding, and can get lodged into your eye. Some standard welding helmets have a flip-up shade and a clear plastic piece underneath, so the helmet can stay down when you chip. Auto-darkening helmets work the same way in that you can see through them when they are down. This will prevent hot slag from burning the rest of your face.

7. Keep sleeves and pant cuffs rolled down and your collar buttoned up. Make sure your clothing has no frays, which can start on fire and cause a serious burn.

8. Use pliers to pick up hot pieces of metal. Hot metal will quickly ruin gloves and can lead to injury. For your own sake and the safety of others, keep track of hot metal and mark it with a soapstone as "hot" when necessary.

Consider the circuit

When you connect the clamp of your welding machine's work lead to a metal table, any area of the table can become part of the welding circuit. If the table is touching another metal object, that object can become an alternative circuit to ground. Do not allow scaffold, steel cable supports, hoist chains, and especially gas cylinders to become part of your welding circuit. Cylinders may explode if damaged, so never strike an arc on a compressed gas cylinder or allow an electrode to touch a cylinder. Also keep cylinders away from sparks and hot metal.

Electrical shock

If you are arc welding in perfectly dry conditions, your risk of being shocked is very minimal. However under certain circumstances, like when water becomes a factor, it is possible to get hurt. One day just after sunrise, a farmer dragged his welding leads across a grassy field to a tractor needing some repairs. He lay down underneath the tractor, dropped his helmet, and struck an arc. At than moment he experienced a sharp jolt of electricity from a low voltage shock that bit hard. The morning dew on the grass was enough moisture to send the electricity on another path. Fortunately for him, this incident tripped the breaker switch in the fuse box, saving his life.

It takes less than one ampere to kill a person. Since the amperage used in arc welding is always considerably higher than that, keep the following safety concerns in mind:

1. Keep gloves, clothing, and feet dry to avoid electrical shock. If there is a chance the glove you are wearing will get wet, have several pairs of gloves handy so you can easily change to a dry pair.

2. Don't weld in the rain.

3. Make sure cables and ground and lead connections are in good condition and free of cracks.

4. Connect your work clamp close to the welding area. This allows the electricity to flow through the base metal and to the work clamp in the shortest possible distance.

5. The risk of electric shock is much greater when using AC.

6. High frequency AC waves used with GTAW power sources can also cause electric shock. Position the work clamp to keep the welding circuit out of the path of your body.

7. Use only properly grounded equipment and always disconnect the power to any arc welding machine *before* changing the leads, opening any protective covers, or servicing the equipment.

8. People with pacemakers are especially vulnerable to welding currents.

Fumes

Avoid overexposure to fumes generated by arc welding, especially with the SMAW and FCAW processes. Keep your head out of the smoke plume to avoid breathing welding fumes. Use enough mechanical ventilation. In some cases, natural air movement from an open garage door or being outdoors will provide enough fresh air. In general, fewer fumes are generated by GTAW, GMAW, and other "wind-sensitive" welding processes. Use a local exhaust system and/or a respirator where necessary. See the fumes section in the safety chapter for more specific information.

This industrial-grade Hobart power source has two levels of amperage adjustment and arc control. If it were being used in production welding, a remote control for amperage could be attached to the machine. *Monte Swann*

Controlling ignition sources

Always remove any flammable or combustible materials from the area before welding or cutting. If you are arc welding on a car, don't forget to disconnect the battery to prevent damage to the car's electrical circuits or a possible explosion. If some materials cannot be removed from the area, use pieces of cardboard, plywood, or sheet metal to deflect the sparks away from the flammable materials. Always know where your sparks are going. When possible, have a friend watch for fire. It may help if someone is ready with a fire extinguisher in case an emergency arises.

A student of mine was working with a friend who was welding outdoors in a grassy area. She made a trip to the work shed for some tools and returned to find her friend surrounded by flames. No fire extinguisher was handy. Her friend kept running the bead, unaware of the inferno. Eventually she got his attention and, before the fire spread to the surrounding trees and buildings, they were able to smother it with a leather jacket.

WELDING MACHINES AND POWER SOURCES

Spending the money to purchase welding equipment is a big financial decision for most people. Each person or shop will have unique requirements, and the best equipment to buy will depend on different factors. Read through the section on purchasing welding equipment and setting up your shop in the safety chapter. Think about which processes will suit your needs and read through the chapters for the specific advantages and disadvantages of each. There is no one piece of equipment that will do it all. But knowing a little about welding will make shopping for machines a lot easier.

The best place to purchase an industrial-quality welding machine is at a local welding supplier. Welding machines sold at big chain stores are usually not industrial grade. Even with brand-name machines, lower-quality, less-powerful versions are sold at hardware and home improvement stores, greatly limiting the size of electrodes you can use. Low-amperage capacity means the machine will work to weld thin-gauge sheet metals only. Small-diameter wire and stick electrodes cost more per pound than the less-expensive, larger-diameter electrodes that can be run on industrial-quality machines.

Avoid purchasing machines with amperage or voltage controls that click into place. There is no possible way to make fine-tuning adjustments to your heat settings without a rheostat control that turns smoothly.

The size of a welding machine is determined by its output capacity and is measured in amps. A machine's amperage is given in the machine's designation. For example, a Dialarc 250 P has a maximum output of 250 amps.

- 150–200-amp machines are used for light-duty welding
- 250–300-amp machines are used in manufacturing shops for production welding
- 400–600-amp machines are used at construction sites for heavy-duty welding and cutting large beams, containers, and pipe

When comparing welding machines, amperage and voltage capacities can both be considered by calculating wattage. The total wattage of a machine is determined by multiplying volts by amps. Volts X Amps = Watts. For example, the Miller 251 wirefeed has an output of 28 volts at 200 amps; $28 \times 200 = 5600$ watts. The Hobart RC-256 wirefeed has an output of 24 volts at 200 amps; $24 \times 200 = 4800$ watts. In this case, at the same amperage, the Miller 251 has a higher output capacity than the Hobart RC-256.

Electrical requirements

Follow the manufacturer's recommendations on the type of input power needed for your welding machine(s). The owner's manual and the data plate on the machine will have information about electrical requirements. Wiring your space for welding is an important consideration. All household power is available at 230 volts, single phase. A 220/230-volt power supply can be used in any home, but the circuit will need to have a certain minimum amperage capability. Larger-diameter wiring is used, along with the appropriate-sized outlet and circuit breaker or fuse. The manufacturer may allow a smaller amperage service for a given voltage, but the maximum output power will be less. If you need 220/230-volt service, you can wire your own home or, if you are not handy with electricity, hire an electrician. However you get it done, have the job inspected before you start welding.

Fortunately, as welding machines have gotten smaller and more portable thanks to inverter technology, the amount

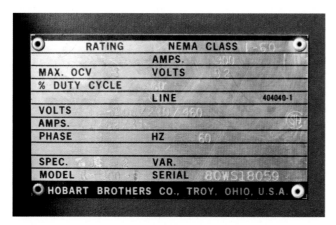

RATING	NEMA CLASS	I-60
	AMPS.	300
MAX. OCV	VOLTS	82
% DUTY CYCLE		60
	LINE	404040-1
VOLTS	200/230/460	
AMPS.		
PHASE	HZ	60
SPEC.	VAR.	
MODEL	SERIAL	80WS18059
HOBART BROTHERS CO., TROY, OHIO, U.S.A.		

Look for the data plate on a welding machine. It will have the important electrical information about the machine. This data plate is for the Hobart Mega-Arc 300, which is a three-phase-only machine. *Monte Swann*

of current required to operate them has also been reduced. There are several good machines on the market that plug into a standard 110/115-volt outlet and have a decent output that can handle larger-diameter electrodes (⅛ inch). These machines will be smaller than those which use 220/230 volts. Still, as good as they might now be, count on only being able to weld limited thicknesses of metal with a machine that runs off a 110/115-volt circuit.

Most new machines can be wired for either single- or three-phase power. All residential homes have single-phase power. Avoid buying a used machine that can only be used with three-phase. You can buy three-phase-to-single-phase converters, but they are very inefficient. Using a phase converter consumes a lot of electricity, and the remaining power may not be sufficient to maintain a welding arc.

When welding at home, remember that your machine's output power may be lower at some times than it is at others because the available voltage to your household power can change. On hot summer days when air conditioners are running and demand for electricity is high, the power company may make adjustments to the available voltage on the electrical grid.

The electrical characteristics for arc-welding machines are broken into three categories: constant current (CC), constant voltage (CV), and a combination of both (CC/CV). The welding process you choose will have specific arc characteristics, so the power source must be capable of producing that type of arc.

Constant current (CC)

CC power sources are used for both SMAW and GTAW processes. This type of power source can provide AC only, direct current DC only, or AC and DC both, depending on the type of machine. Remember that current is synonymous with amperage. A CC machine supplies a constant amount of

amperage to the welding arc, even when there is a large change in voltage. The relationship between voltage and amperage can be plotted on a curve and is generally referred to in volts changed per 100 amps. The shape of a machine's curve determines the characteristics of its welding arc. The curve for CC is a steep line. That is why constant current power is known as a drooper.

This Miller 110A CC power supply comes with a 220-amp plug. An adapter was made with standard hardware store items so the machine can be plugged into a standard household 110-amp outlet. In order to do this safely, the power source must have voltage-sensing capabilities. Check with the manufacturers recommendations to find out if a machine is voltage-sensing. *Monte Swann*

This Miller Dialarc 250 P is a transformer-rectifier, AC/DC, single-phase, CC power supply with a duty cycle of 30–50 percent. This SMAW power source can be set up for GTAW welding with DC polarity. The maximum output is 250 amps. The input can be wired for 220/230 volts. *Monte Swann*

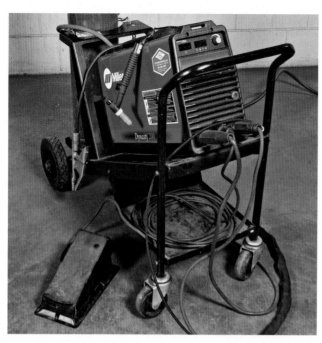

This Miller Dynasty 200 is a CC inverter, AC/DC power supply that can be set to either single-phase or three-phase power. This GTAW power source can be set up for use with either 110/115-volt or 220/230-volt input power. *Monte Swann*

This Miller Dynasty 300 CC inverter AC/DC TIG welder, which is capable of handling any TIG or stick-welding job, can only be used with 220/230-volt input power. Below the machine is the coolant pump that cools the TIG welding torch. *Monte Swann*

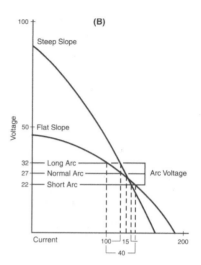

A typical constant current Volt-Amp curve. *Welding Inspection Technology, Fourth Edition, Figure 3.7b, Reproduced with permission from the American Welding Society (AWS), Miami, FL USA*

Some CC machines are designed specifically for SMAW, while others are designed for GTAW. There are, however, ways to make one machine work for both processes. A machine designed for GTAW will only need an electrode holder, instead of a TIG torch, in order to perform SMAW.

Although an SMAW machine won't have controls such as high-frequency arc start or a remote amperage control, it can still be used for GTAW. Machines designed for SMAW will need a few accessories before GTAW can be performed. Purchase a valved air-cooled TIG torch, regulator flow meter, and argon cylinder. Attach the welding lead of the TIG torch to the negative terminal of the SMAW machine and use the adjustment knob on the torch to turn the shielding gas on and off. The electrode will have to be scratch started against the work to start the arc. Once you begin welding, the amperage on the machine is set and cannot be changed while the arc is on, but good-quality TIG welds can be made.

Constant voltage (CV)

Any of the wirefeed welding processes will require a CV power source. CV, also known as constant potential, is used in welding with both solid and tubular wires in GMAW and FCAW. This

Look for the gas inlet in the back of any wirefeed welding machine to see if it can be used for GMAW. Gasless wirefeed machines without a gas solenoid can only be used with self-shielded tubular wires (FCAW-S). *Monte Swann*

type of machine will hold a narrow voltage range, even through changes in arc length and amperage. The voltage remains constant while the amperage can vary. When plotted out on a graph, the volt/amp curve of a constant voltage power source has a shallow angle, which is opposite of the constant current power. See the GMAW chapter for more information on the relationship between amperage and voltage with these power sources.

CC/CV machines

This type of power source provides the most welding versatility. Machines with both CC and CV can be used with any of the arc-welding processes discussed in this book. The flexibility comes at a higher price. These machines are at the top of the price range and require an additional wirefeeder unit to perform GMAW and FCAW. Wirefeeder units are usually sold separately.

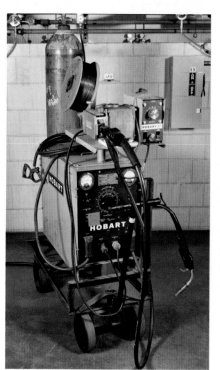

This example is the Lincoln SP 125 Plus, which is a CV, single-phase, 110/115-volt input power machine with a maximum output of 18 to 20 volts and a 20 percent duty cycle. This small, industrial-grade wirefeed welding machine works for GMAW short-circuit transfer only, with wire sizes up to 0.030 inch. The Lincoln SP125 can also be used for FCAW-S with a 0.035-inch-diameter wire. *Monte Swann*

This Miller 251 is a CV, single-phase, 220/230-volt input power machine with a maximum output of 28 volts and a 40–60 percent duty cycle. This midsized wirefeed welding machine can be used for GMAW or FCAW with 0.023-0.045-inch wire diameters. A spool gun can be purchased separately that allows you to weld stainless steel and aluminum. *Monte Swann*

This Hobart is a CV, three-phase-only machine with a maximum output of 24 volts and a 100 percent duty cycle. Although this wirefeed machine will do a good job welding steel and stainless steel with GMAW and FCAW, the three-phase electrical input requirement limits its use for home shops. *Monte Swann*

This Miller Axcess is a CV single- or three-phase machine that requires 220/230-volt input power, has a maximum output of 44 volts, and a 100 percent duty cycle. This large welding machine is commonly used for production fabrication in industrial welding shops. The Miller Axis has GMAW pulse-spray transfer capabilities, and can be used with a wide range of wire diameters, from 0.030 to 0.062 inch. *Monte Swann*

INTRODUCTION TO ARC WELDING

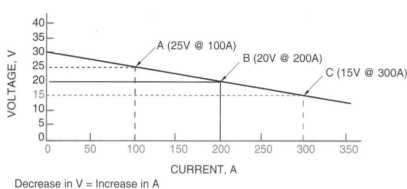

A typical constant voltage (constant potential) Volt-Amp curve. *Welding Inspection Technology, Fourth Edition, Figure 3.13, Reproduced with permission from the American Welding Society (AWS), Miami, FL USA*

Decrease in V = Increase in A

Duty cycle

Most machines are not designed to provide power for arc welding 100 percent of the time. In most cases, people do not have their welding arc on 100 percent of the time either. A machine's duty cycle is measured in the amount of time the arc can be on in a 10-minute time span. A 60 percent duty cycle means you can weld with the machine for 6 minutes out of every 10.

As a machine's size increases, so does the duty cycle and price. You'll be surprised how long 6 minutes of continuous welding actually is. Even a duty cycle of 2 minutes out of every 10 (20 percent) is adequate for most home shop projects. If you do exceed a machine's duty cycle rating, the power source should automatically shut off to prevent burning out the interior components.

Open circuit voltage (OCV)

Open circuit voltage (OCV) is the amount of voltage available when the machine is turned on and no welding is being done. OCV varies between 50 and 100 volts. When an arc is struck and welding begins, the voltage drops due to the resistance created by the arc gap. Arc voltage (working voltage) is lower than OCV, usually between 18 and 36 volts. This low voltage combined with high amperages (producing the heat required for welding) is not particularly dangerous if proper grounding and insulators are used in a dry environment.

Machine types

There are different types of welding machines designed to supply CC, CV, or CC/CV power:

- *Transformer:* This type of CC machine will only produce AC. This is the base line model welding machine for SMAW because it is the least expensive to purchase and takes less electricity to run. Commonly called welding transformers, the machines run on single-phase power. Lincoln Electric named their version of the machine "buzz box" and Miller named it the "thunderbolt." In terms of electrical equipment, a transformer changes the voltage of AC from one level to another.

- *Transformer-rectifier:* This type of power supply is available in CC, CV, or CC/CV types. A welding rectifier has a transformer coupled with a rectifier, the device that changes AC into DC. Transformer-rectifiers can produce AC and DC, although DC requires more electricity to weld at the same amperage as AC. This type of older conventional machine is larger and heavier, requires more input power, and uses much more electricity than an inverter. The machine will have to be physically rewired depending on the input power being used. Most transformer-rectifiers have dual current controls: a range switch or rough adjustment and a rheostat dial for fine adjustment.

- *Inverter:* An inverter changes DC back into high-frequency AC. They can be CC, CV, or CC/CV types. Inverters have gained popularity because the machines are smaller (75 percent smaller than conventional machines), are lighter weight, take less input power to run (many run on 110/115 input power), and use much less electricity than transformer-rectifiers. Inverters are more expensive, but the advantages are worth the extra cost. Most inverters are equipped with voltage sensing, meaning that the machine analyzes the input power available and makes adjustments accordingly. There is no need to change the wiring in the machine; if not enough power is available, the machine will show a fault and not run. Inverter-type controls can produce variable frequency for fine-tune adjustment of the arc.

- *Engine-driven:* When there is no source of power available on a job site for arc welding, an engine-driven generator (running on gas, diesel, or propane fuel) is used to provide the electric current. Common versions of this machine produce only DC and some produce AC or DC, having excellent arc characteristics. These machines need proper routine maintenance and, in addition to welding fumes, the exhaust from the welding machine can be hazardous. Engine-driven machines can be very loud and are designed for outdoor use. The cost of electricity for plug-in machines is less than the typical fuel costs for an engine-driven machine, but the plug limits their portability.

CHAPTER 10
SHIELDED METAL ARC WELDING (SMAW)

It is estimated that about half of all the welding done in the United States is with the SMAW process. No matter how big or small your project is, SMAW can be used with good results, but only if you have the necessary knowledge and skill set.

Many people shy away from the SMAW process because they think it is too difficult to learn. It will require a little more time and practice to become proficient at SMAW. The payoff is being able to utilize a more versatile process than the alternative of "gasless" wirefeed, which requires less manual skill. Let's take a look at each part of the acronym SMAW:

Shielded

When you think of a shield in general terms, it is something used for protection. Welds need to be protected from the atmosphere, which contains nitrogen, oxygen, and hydrogen. Molten metal will rapidly absorb these elements if left unprotected, making the weld more brittle. The shield in SMAW is the flux coating on the outside of the electrode (welding rod), which creates CO_2 when burned. There are many types of metal electrodes and flux coatings.

Metal

At the center of the welding rod, mostly covered by the flux coating, is the filler metal. Electrodes will have different filler metal compositions, depending on the type of base metal to be welded. Carbon steels, copper alloys, stainless steels,

and cast iron can all be welded with the SMAW process. Low-alloy steels are generally welded with the low-hydrogen electrode group.

Arc

When electricity crosses an open gap, an arc is created. The welding arc creates resistance in the electrical circuit. Resistance generates the heat required to melt the base metal. The filler metal from the electrode travels across the arc stream and is deposited into the molten pool.

Welding

A weld bead is created when both the base metal and filler metal become molten, fuse together, and solidify.

MACHINE CONTROLS

One advantage or disadvantage, depending on your viewpoint, of SMAW is the small number of machine settings and adjustments that need to be made. Beginners may find that navigating the small number of controls simplifies things, whereas advanced welders may want more control over the characteristics of the welding arc. Although there are fewer machine controls to consider with SMAW, each one plays an important role in successful welding.

A CC power source will be required for SMAW (see chapter 9 for more information on these types of welding machines). Welding electrodes may require AC or DC, or some may be used with either type of current. Depending on the application, either straight or reverse polarity may need to be used. Know how to switch current on the machine. Polarity can be changed by a button or switch on some power sources. Others require that you physically change the leads to different terminals, with a quick-connect or lug-type connection.

With SMAW, the level of amperage determines the heat for welding. Amperage is set directly on the machine, usually with a dial. On a machine with a digital readout, exact amperage settings can be made. On machines without a digital display, use the numbers around the dial for your reference. Some rheostat dials will have a certain amount of play in them. To make consistent changes to amperage, turn the dial down past the setting you need, and then turn it back up to the required setting. If you just turn the dial to a lower amperage, the slack in the dial could account for a big difference in the amount of current.

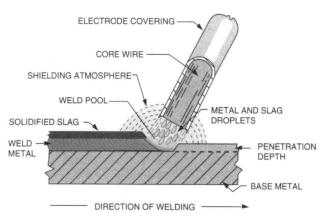

Welding Inspection Technology, Fourth Edition, Figure 3.2, Reproduced with permission from the American Welding Society (AWS), Miami, FL USA

This AC/DC CC power supply has a rated capacity of 300 amps, which is more than enough for most welding applications. It also has the advantage of readout dials for both amperage and voltage. *Monte Swann*

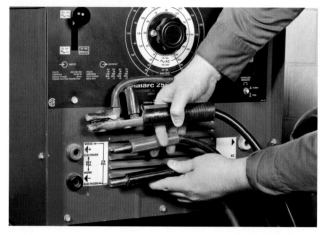

In these exercises, the electrode holder is connected to the positive terminal of the welding machine, and the work clamp to the negative for DCRP. *Monte Swann*

Arc control

Machines equipped with an arc control feature can be fine-tuned. Also called arc force or slope control, this feature changes the characteristics of the welding arc. A low setting decreases the amount of current in relationship to voltage. The arc is quiet, with less penetration into the base metal—a good setting for welding sheet metals less than 3/16-inch thick.

Turning up the arc control to maximum will flatten the volt-amp curve, increasing the amount of current in relationship to voltage. The result is an arc that will penetrate deeper and works well with thicker metals. Changing your arc force will only fine-tune the arc and is not effective when an incorrect amount of amperage is used.

Amperage

Amperage provides the heat needed for welding. Too little amperage, and there won't be enough heat; too much amperage and there will be too much heat. Fusion will result only when the base metal has been melted and

liquefied filler metal from the electrode flows readily into the molten pool. But, it is possible to overheat the base metal, filler metal, or both. Unfortunately, there is no exact setting for amperage. Handheld amperage calculators come in handy, or look for suggested amperage settings from the electrode manufacturer.

One of the best ways to learn about amperage settings is to cover all the numbers on the machine's amperage selector dial or display with tape and just start welding. Using the same metal thickness, weld a little, change the dial, weld some more, change the dial, and so on. Then change metal thickness and repeat the process. Do this as many times as you can. This will force you to take a close look at the effects of different amperages and enable you to learn to read the molten puddle faster.

Too little amperage will make it difficult or impossible to strike an arc; if you do get an arc, the welding pool will be small. More amperage will increase the depth of fusion into the base metal. But too much will make the weld pool very fluid and hard to control. Excessive amperage can also boil the liquid metal and burn the flux on the electrode. I've even seen the flux catch on fire.

It is important to remember that these amperage settings are only a good place to start. The final amperage setting is determined by the following factors:

- *Metal thickness:* Thicker metals will dissipate the heat, requiring more amperage. Thinner metals require less amperage and smaller-diameter electrodes.

Guidelines for Setting Amperage to Start

E7018/7014

Diameter Amperage	Decimal Equivalent	Beginning
3/32 inch	.093	93 amps
1/8 inch	.125	125 amps
5/32 inch	.156	156 amps

E6010/6011 — use the same conversion as above only subtract 1/32 of diameter

3/32 inch (−1/32 = 1/16 inch)	1/16 = .063	63 amps
1/8 inch (−1/32 = 3/32 inch)	3/32 = .093	93 amps
5/32 inch (−1/32 = 1/8 inch)	1/8 = .125	125 amps

E6013 — use slightly less amperage than what is recommended for 7018 (decrease amperage by about 10 percent)

1/16 inch	.063 − 10%	57 amps
3/32 inch	.093 − 10%	84 amps
1/8 inch	.125 − 10%	112 amps

This Miller power source has four sets of numbers around the outside of the amperage dial. One set is specific to the type of current being used (AC or DC). The other set is for amperage on the high range or low range, depending on where the rough adjustment is set. Be sure to read the correct set of numbers when using a machine like this. *Monte Swann*

- *Position of weld:* Welds made in the vertical position require lower amperages to maintain the molten weld pool.
- *Joint design:* Butt and corner joints require less amperage than tee and lap joints. Fusion faces in close proximity to each other require higher amperages.
- *Welder's ability:* Beginning welders should use smaller-diameter electrodes and lower amperages because the smaller molten puddle is easier to carry. More experienced welders will use faster travel speeds and more advanced techniques to carry more heat.

Always remember that machines run at different amperages, depending on the age of the machine and how it was manufactured. A setting of 100 amps on one machine can be very different from the same setting on another. Variations in the windings of different machines, different electrical connections, input power, and/or the scaling of amperage by different manufacturers all account for differences in actual amperage. A volt/amp meter connected to the electrode holder while the arc is on and welding is being done is the only accurate method for determining actual amperage.

The most important factor is being able to read what the molten puddle is doing and make adjustments as necessary. This ability comes with experimentation, practice, and experience.

ADVANTAGES

The SMAW process is used in a variety of applications from welding under water to high beams of a skyscraper rising above a city. The shielding atmosphere created by the flux coating on the electrode (welding rod) is fairly wind resistant allowing SMAW welding practically anywhere. There are few parts to the equipment: a power source, two welding leads, a work clamp, and an electrode holder. This simple setup is easy to maintain, very portable, and longer lead lengths can be used to weld hard-to-reach places.

It is possible to burn through a plate and cut metals apart with an SMAW electrode. To do this, use amperage twice as high than what is used for welding. *Monte Swann*

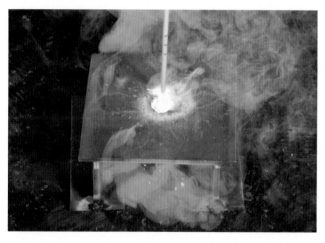

Circle the electrode and push down until a hole appears. Keep the end of the electrode at the same height as the bottom side of the plate. *Monte Swann*

Circle the electrode to widen the hole. Notice there will be a large amount of fumes generated during this process. *Monte Swann*

Recommended Copper Welding Cable Sizes

Power Source		Awg Cable Size for Combined Length of Electrode and Ground Cables				
Size in Amperes	Duty Cycle, %	0 to 50 ft (0 to 15 m)	50 to 100 ft (15 to 30 m)	100 to 150 ft (30 to 46 m)	150 to 200 ft (46 to 61 m)	200 to 250 ft (61 to 76 m)
100	20	6	4	3	2	1
180	20-30	4	4	3	2	1
200	60	2	2	2	1	1/0
200	50	3	3	2	1	1/0
250	30	3	3	2	1	1/0
300	60	1/0	1/0	1/0	2/0	3/0
400	60	2/0	2/0	2/0	3/0	4/0
500	60	2/0	2/0	3/0	3/0	4/0
600	60	2/0	2/0	3/0	4/0	*

* Use two 3/0 cables in parallel.

Welding Handbook Volume 2, Eighth Edition, Table 2.1, Reproduced with permission from the American Welding Society (AWS), Miami, FL USA

A wide range of metal thicknesses can be joined with SMAW, with the least amount of changes to the welding machine and setup. You should use 1/16-inch diameter electrodes on thin-gauge sheet metals while 1/8-inch diameter can successfully weld materials over 1-inch thick. All different sizes—1/16 inch, 3/32 inch, 1/8 inch, and even 5/32 inch—can be used with a DC output of 150 amps.

DISADVANTAGES

A flux-coated electrode is always used in SMAW, so there will always be slag covering the finished weld bead. Slag inclusions can easily occur in a welded joint if improper technique is used. Some ingredients in the flux are toxic. Inhaling the fumes can cause dizziness, nausea, irritation, or dryness of eyes, nose, and throat. It may also worsen asthma and emphysema if you already have those conditions. Long-term exposure can lead to more serious health problems. Although aluminum electrodes were developed for SMAW, the process is not suitable for welding aluminum.

EQUIPMENT
Leads (welding cables)

Insulated copper cables, or leads, come in a variety of sizes based upon the diameter. Since welding leads carry the electricity to the work clamp and electrode holder, they need to be the proper size to conduct the welding current efficiently. The size needed depends upon the amount of amperage used, the maximum output capacity of the machine, and the length of the leads. Large-diameter leads are needed for longer lengths, because there is more resistance in the circuit. The large diameter gives the amperage more area to flow through, reducing the amount of resistance. Keep your leads in good condition. Cracked and stripped off insulation can increase the chances of electric shock, especially if water is present. The copper wires in the cable are very fine and can be crushed or kinked, creating a bottleneck in current flow. So will using leads that are too small to handle the length of wire or amperage. These circumstances will cause a large buildup of resistance in the circuit, overheating the equipment, and possibly starting a fire.

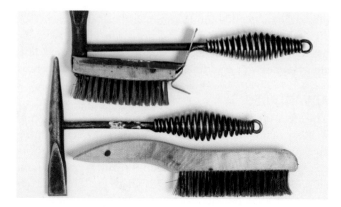

Tweco and Lenco brand electrode holders. The ground clamp on the right is made of brass and has a 300-amp rating. In between the two electrodes is an insulator and copper part of the electrode clamp. *Monte Swann*

A chipping hammer and wire brush can be purchased separately or as a combined tool. Notice the combined tool does not have the pointed end of the standard chipping hammer. *Monte Swann*

Electrode holder (stinger)

Like welding leads, electrode holders are sized depending upon the amperages used. If your machine is set for welding with 210 amps, it is safer to use a 300-amp-rated electrode holder than one rated for 200 amps. If the electrode holder becomes too hot to hold, or the end of your copper cables look discolored, check for problems. Either the electrode holder is underrated or the welding leads are too long or too small.

Insulators on the outside of the electrode holder prevent arcing when it is set down on a metal worktable. Each insulating piece should be replaced when worn. If an uninsulated part of the stinger makes electrical contact, it can cause damage to the equipment, electrical shock, or an explosion. The handle of an electrode holder is spring-loaded and can clamp onto a welding rod. This spring-loaded release lever is helpful when you need to let go of an electrode quickly, like when it is stuck to the work piece. Copper teeth, protected by insulators, hold the bare round end of the electrode in place when welding. Grooves machined into the copper allow the electrode to be clamped at different angles for different welding positions and for more comfortable arm and body positions. With some electrode holders, the copper teeth are replaceable when they wear out. When finished welding, always remove the electrode from the holder to avoid unintentional arcs.

Work clamp (ground clamp)

When using reverse polarity, the electricity goes through the work clamp first, then travels through a work table (if there is one in the circuit), through the base metal, and to the welding arc. Work clamps come in different sizes and are rated by amperage. Since they are an integral part of the welding circuit—no matter what type of current or polarity is being used—never use a ground clamp at higher amperages than what it is rated for. Connect your work clamp to bare metal. If the base metal is rusty, dirty, corroded, or painted, sand or brush off an area of bare metal for the work clamp. Otherwise, contaminants between your clamp and base metal will cause resistance in the circuit, impede the amperage flow, cause heat buildup, and destabilize the welding arc.

Work clamps are made that bolt onto the base metal. Some welders use a scrap piece of metal for a work clamp. The welding lead is bolted onto the scrap piece. Next, the scrap piece is welded directly onto the base metal, creating the best possible connection. Other welders will lay the work clamp on the table or the metal being welded. Although this method will work, it is not a good practice since the clamp can easily be bumped or pulled off the metal; small areas of contact cause resistance in the circuit. Always connect your work clamp securely to the table or base metal.

Electrodes

Known as welding rods or sticks, electrodes have a metal core with a flux coating on the outside. SMAW uses a consumable electrode that becomes part of the finished weld bead. Both the composition of the filler metal and flux will have an impact on the shape, strength, and mechanical properties of the finished weld. Different systems of identification are used to organize electrodes. Most major manufacturers will have specific names or designations for their different electrodes, but also use the AWS system of electrode classification. For example, E6010 is the AWS classification and 5P+ is the manufacture's designation for the E6010 5P+ welding electrode. Usually, the letters and numbers are printed on the flux coating of every electrode. There are other companies that use their own system. In any case, consult the manufacturer for information on electrodes you are using.

The type and size of electrodes from top to bottom are:
- Eutectic X-Tron 244 nickel alloy, probably used for welding cast iron
- ⅛-inch-diameter 309-16 stainless steel
- Old copper alloy electrode from an unknown manufacturer. Color coding dots were used to identify electrodes before numbers were printed on them.
- Hardalloy 32 hard-facing electrode used for building up surfaces to withstand high impact loads.
- ⅛-inch-diameter 6010
- ⅛-inch-diameter 6011
- ³⁄₃₂-inch-diameter 6011
- ⁵⁄₃₂-inch-diameter 6013
- ⅛-inch-diameter 6013 Lincoln fleet weld 57
- ³⁄₃₂-inch-diameter 6013
- 7018 H4R ESAB atom arc

Monte Swann

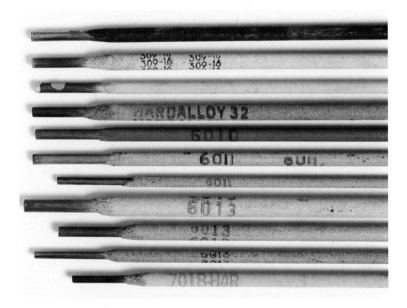

F-No. Classification	Current	Arc	Penetration	Covering & Slag	Iron Powder
F-3 EXX10	DCEP	Digging	Deep	Cellulose-sodium	0–10%
F-3 EXXX1	AC & DCEP	Digging	Deep	Cellulose-potassium	0%
F-2 EXXX2	AC & DCEN	Medium	Medium	Rutile-sodium	0–10%
F-2 EXXX3	AC & DC	Light	Light	Rutile-potassium	0–10%
F-2 EXXX4	AC & DC	Light	Light	Rutile-iron powder	25–40%
F-4 EXXX5	DCEP	Medium	Medium	Low hydrogen-sodium	0%
F-4 EXXX6	AC or DCEP	Medium	Medium	Low hydrogen-potassium	0%
F-4 EXXX8	AC or DCEP	Medium	Medium	Low hydrogen-iron powder	25–45%
F-1 EXX20	AC or DC	Medium	Medium	Iron oxide-sodium	0%
F-1 EXX24	AC or DC	Light	Light	Rutile-iron powder	50%
F-1 EXX27	AC or DC	Medium	Medium	Iron oxide-iron powder	50%
F-1 EXX28	AC or DCEP	Medium	Medium	Low hydrogen-iron powder	50%

Note: Iron powder percentage is based on weight of the covering.

Welding Inspection Technology, Fourth Edition, Figure 3.4, Reproduced with permission from the American Welding Society (AWS), Miami, FL USA

Flux

The flux coating on the outside of an electrode has several functions. The flux:

1. Acts as a cleansing and deoxidizing agent in the molten weld pool.
2. Creates an inert atmosphere by releasing CO_2, shielding the molten pool from contamination.
3. Forms a slag coating over the semi-molten metal to further protect it from the atmosphere.
4. Stabilizes the welding arc.
5. Improves penetration (depth of fusion) into the base metal.
6. Determines type of welding current used for welding: AC, DCEP, or DCEN.

The chemical composition of the flux has a great impact on the finished weld bead. In the mild-steel electrode group there are three main types: clay-based (rutile), paper-based (cellulous), and limestone (potassium carbonate), also know as low hydrogen. Each has its own application in welding.

AWS ELECTRODE NUMBERING SYSTEM

- All mild-steel and low-alloy steel electrodes are classified with a four- or five-digit number prefixed by the letter E, which stands for electrode.
- The first two or three digits give the tensile strength in kilo-pounds per square inch (ksi). An E6011 electrode has a tensile strength of 60,000 psi and an E11018 has a tensile strength of 110,000 psi. This given strength assumes that all welds are made properly.
- The third (or fourth) digit indicates the weld positions the electrode can be used: 1 = all positions (flat, horizontal, vertical, and overhead); 2 = flat-position groove welds (flat and horizontal fillet welds only);

3 = flat position only (this designation is no longer used). 4 = flat, horizontal and vertical downward progression only.

- The last digit indicates the type (chemical composition) of flux coating.
- A suffix is sometimes used at the end of the four- or five-digit classification to designate the composition of the deposited weld metal. For example, the suffix A1 indicates the electrode contains 0.12 percent carbon, 0.06-1.00 percent manganese, 0.40-0.80 percent silicon, and 0.40-0.65 percent molybdenum.
- Stainless steel electrodes are identified by their AISI classification number. See the chapters on types of metals for more information.
- Alloy electrodes are identified in a similar way to brazing and soldering filler alloys. Each alloy is listed by a letter designation. For example, NiCr means the electrode contains nickel and chromium.
- Be aware, some electrode manufacturers have their own system of identification. When in doubt contact the manufacturer for specific information about the electrode including its uses, mechanical properties, alloy content, flux composition and storage requirements.

Refer to chart 10-1 on page 114 for more information on mild-steel and low-alloy steel electrodes. Remember, these numbers do not indicate electrode sizes, nor are electrode sizes printed anywhere on the electrode itself. Use a caliper to measure the bare metal end to determine an electrode's actual size.

SELECTING THE CORRECT ELECTRODE
Properties of the base metal

The manufacturers of electrodes have done a considerable amount of research in developing their products. Electrodes are specifically designed to weld certain base metals, such as carbon steels, low-alloy steels, and stainless steels. Before you begin a project, the first thing to know is the type of base metal being used. Then, select an electrode recommended for that metal.

Even the most experienced welder will never be expected to know everything about the wide variety of electrodes available today. Asking a welding supplier or doing a little research can save a lot of time and money and can provide peace of mind. There are those people who will pick up any old rod and start welding with it. Rod burners may get lucky sometimes, but are not the same as serious welders who will take the time to get things right.

Electrode diameter

An electrode diameter equal to about half the thickness of the base metal will make the heat input more controllable. Never use an electrode larger than the thickness of the base metal. When welding vertical and overhead, it may be necessary to use smaller electrodes; 3/16 inch is the largest diameter used for these positions regardless of plate thickness. Joint design can also be a factor. On thick metal with a narrow V-groove, you may need a smaller-diameter electrode to ensure penetration at the root of the weld.

Joint design and fit-up

Groove joints with insufficient beveled edges or open gaps require deep penetrating and fast freezing electrodes for the first pass. Fillet welds can be entirely welded with a more fluid, light-penetrating electrode with a slower rate of solidification.

Welding position

Check the third (or fourth) digit of the electrode to find out if the electrode can be used in all positions. If you are making a vertical weld, using a fast-fill electrode like E7024 is not an option.

Welding current

Using an AC-only machine will limit the number of choices you have in electrodes. AC current travels in both directions due to the positive and negative poles constantly reversing, creating an amperage lag. Some electrodes have additional stabilizers added to the flux, designed to maintain the arc through the lag. Other electrodes are designed to run only with DC. A power source with both AC and DC will allow you to use any electrode.

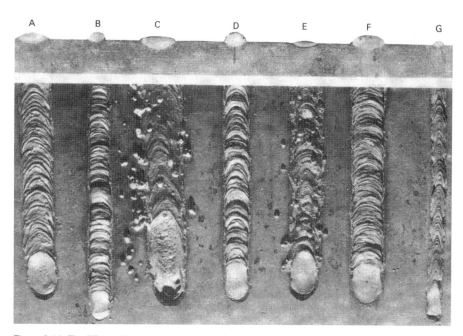

Figure 2.11—The Effect of Welding Amperage, Arc Length, and Travel Speed; (A) Proper Amperage, Arc Length, and Travel Speed; (B) Amperage Too Low; (C) Amperage Too High; (D) Arc Length Too Short; (E) Arc Length Too Long; (F) Travel Speed Too Slow; (G) Travel Speed Too Fast *Welding Handbook Volume 2, Eighth Edition, Figure 2.11, Reproduced with permission from the American Welding Society (AWS), Miami, FL USA*

To get into tight spots or make sharp corners in a weld, an electrode can be bent at an angle. Weld only until you reach the end of the flux coating. *Monte Swann*

WELDING TECHNIQUE

Striking an arc

Get your electrode as close to the beginning of your weld as possible, without touching the metal. Lower your helmet and lightly drag the tip of your electrode across the work piece. When the arc is started, lift the electrode slightly and move it to where you want to begin. If you pull up too far, the arc will go out. If the electrode sticks, either twist it free or let go of it by opening the jaws of the stinger. If you are having trouble striking an arc, it may help to turn up your amperage a little. I usually tell people to strike it like a match.

Chart 10-1

Electrode Number	Characteristics and Uses
Mild-Steel Electrodes	
E6013	General purpose, easy to use, smooth bead appearance Low arc penetration, can be used on sheet metal with DCEN
E6012	For fillet welds on sheet metal; rarely used today Can weld joints with poor fit-up
E6010	Deep penetration with fast-freeze characteristics (quickly solidifies) Minimum amount of base metal preparation required Can weld through light to medium amounts of rust, dirt, and paint Fills gaps; good for first pass in joints with open roots Rough bead appearance with little slag
E6010 5P+	Electrode designed by Lincoln Electric Company Easy arc starting and slag removal Improved bead appearance
E6011	Same characteristic and uses as 6010 except: Slightly less penetration than 6010 Can be used with both AC and DC
E7014	Same characteristics as 6013 except: Iron powder added to the flux for higher deposition rates
Low-Hydrogen Electrodes	
E7016	Older version of the 7018 electrode with a lower deposition rate

Electrode Number	Characteristics and Uses
E7018	Outstanding crack resistance Used to join a wider range of carbon steels than any other electrode Can be used to weld sulfur steels and low-alloy/high-strength steels Smooth bead appearance Must be stored properly, protected from moisture in air
E7018 H4R	H4 = low-hydrogen rating of 4 ml/100 grams R = moisture-resistant coating, not waterproof Also requires proper storage
E7018 AC	Additional stabilizers added to the flux for welding with AC
E7048	Specifically designed for welding joints vertical down Notice the 4 as the welding position
Fast-Fill Electrodes	
E6027	Contains 50 percent iron powder in flux; high rate of deposition Medium penetration, slightly concave bead profile For single-pass welds only
E7024	Low arc penetration No need to maintain arc length, electrode can contact base metal Deposits large quantities of filler metal on thick sections Flux contains 50 percent iron powder by weight Used in production before wirefeed

Push vs. pull

A push or pull method can be used when welding. Pushing and pulling are similar to forehand and backhand techniques in gas welding. Pushing the electrode is traveling in the same direction the electrode is pointed. Pushing will result in slightly less penetration into the base metal and more preheat in the joint. Pulling (dragging) the electrode is moving in the opposite direction the electrode is pointing. Pulling will result in slightly more penetration into the base metal with no preheat in the joint being welded. Try both methods when learning how to weld. Either way will work, as long as the proper electrode angles are maintained.

Starting a bead

Problems with weld beads and discontinuities usually happen at the beginning and/or end of a weld bead. A common mistake made at the beginning of a bead is traveling too fast. The base metal is at room temperature—the coldest it will ever be during the weld. The base metal needs time to heat up so complete fusion can take place. Moving quickly at the very beginning after the arc is struck will prevent the filler metal from penetrating into the base metal, leading to a void in the weld bead, usually filled with slag. When starting a bead, circle the rod a few times to heat up the base metal and get the puddle wide enough.

Electrode Number	Characteristics and Uses
E7028-H8	High deposition with low-hydrogen rating of 8 ml/100 grams Good notch toughness down to 0 degrees F Alloy-steel/low-hydrogen electrodes
E7018-A1	For welding alloy steels with 0.50 percent molybdenum
E8018-C3	Contains 1 percent nickel For repairs on a wide variety of low-alloy and carbon steels with a tensile strength between 70,000 and 80,000 psi Good notch toughness at temperatures as low as -40 degrees F
E9018-M	For making fillet welds on high-strength steels with a tensile strength of 90,000 psi or higher (like T-1, HY80, and HY90)
E11018-M	For all applications of joint designs; both groove and fillet welds on high-strength steels Good notch toughness at low temperatures; high impact resistance Physical properties of the joint meet or exceed the base metal in the as-welded or stress-relieved condition
E4130/4140/4341	All position, low-hydrogen electrode with iron powder in flux For welding low-alloy/high-strength steels and steel castings of the same or similar alloy composition
Stainless-Steel Electrodes	
E308L	Used to weld 302, 303, 304, and 308 stainless steels L = low-carbon content (same for all 300 series electrodes)
E309L	For welding mild steel to stainless steel

Electrode Number	Characteristics and Uses
E316L	For welding 316 stainless steel
E383	For welding 31 percent Ni, 27 percent Cr, 3.5 percent molybdenum steels, like alloy 28
3XX-15 (suffix only)	Electrode used with DCEP only
3XX-16	Electrode used with DCEP or AC Flat to convex bead profile for welds made in horizontal position
3XX-17	Electrode used with DCEP or AC Flat to concave bead profile for welds made in horizontal position Must use slight weave motion when welding vertical up Electrode can be dragged (used with pull technique)
Alloy Electrodes	
Ni 55/ Ni 98	Contains 55 percent and 98 percent nickel, respectively Used for welding cast iron
Ni Cr Fe-2	Contains nickel, chromium, and iron For welding dissimilar metals and nickel-based alloys that require post-weld heat treatment
Hard facing	Used for building up surfaces on parts such as bucket teeth, gear teeth, steel shafts, crane wheels, and rock crushers Electrodes are rated on level of resistance to abrasion and impact, machine ability, and maximum number of layers Should not be used for joining metals
Copper based	Electrode contains copper alloy with silicon, tin, nickel, or aluminum Used for welding copper alloys of the same or similar composition

Restarts

In SMAW, the electrode will deposit only 6 to 10 inches of weld before the rod is totally consumed. Learning to restart and tie back into the end of a previous bead (crater) is a good skill to master. The technique for restarts is the same for SMAW, GMAW, and FCAW.

1. Chip off and brush away any slag at the end of a weld (this step is not necessary in GMAW).
2. Strike the arc in the joint ½ inch directly ahead of where the last bead ended, or see the 3-second rule below.

3. Bring the electrode back to the crater, circle once or twice to fill it in, and continue welding.

In SMAW it will take about 3 seconds for the flux to get hot and begin burning off properly. Restarting the weld directly over the previous crater will not allow time for the flux to start burning, leaving porosity and/or incomplete fusion. A good restart will look like one continuous bead. When you can no longer tell where the restart begins you have done everything correctly.

Stopping

When finishing a bead, a little extra filler metal should be added to prevent leaving a concave crater. At the end of a weld, circle the electrode two or three times, then rotate the electrode up over the bead in a quick and decisive motion to pull out. The electrode is shortest at the end of a weld. Beginners tend to hold the electrode too far away from the base metal (called long-arcing). The end of a weld is the worst place for long-arcing because the metal is at its highest temperature from welding. A long arc length will widen the area of the arc, undercutting the base metal; this can ruin the end of a joint.

DASH

The four main factors in welding technique are distance, angle, speed, and heat (DASH). If you get these four things right, you will have a good weld every time. Keep these in mind when practicing your welds. Experiment by altering one variable at a time and observe what happens when you use different angles or change only the amperage. Your chances for success are far greater if you can consistently control each component of DASH at the same time.

Distance

In SMAW, distance is your arc length (arc gap). The proper distance is a fixed variable in relationship to heat. Using the correct arc length maintains the balance between amps and volts and allows complete fusion to take place. The tip of your electrode should always be a distance slightly less than the diameter of the electrode away from the metal. For a ⅛-inch-diameter electrode, an arc gap of a little less than ⅛ inch is ideal. Maintaining this distance throughout the weld is difficult because as the electrode burns, it is being consumed, getting shorter. Often, beginners are too far away from the base metal; move closer with the electrode to maintain the proper arc length.

Arc length too *short:*

Electrodes are designed to run at a certain amperage and voltage. If the arc length is too short, the voltage decreases and amperage slightly increases. There is less resistance at the shorter arc and less heat provided to melt the base metal. Most of the weld metal will end up sitting on top of the base metal, not penetrating into it.

Arc length too *long:*

As arc length increases, voltage increases and amperage slightly decreases. The arc spreads out as it is forced to jump across a greater distance. The arc becomes erratic and the filler metal begins to form droplets in the arc stream.

Angle

There are two angles to consider. Travel angle is the angle the electrode is held in line with the weld axis (length). Work angle is the angle of the electrode perpendicular to the weld

This is a close-up of how far away your electrode should be from the base metal. Learn to maintain this distance while the electrode is being consumed. *Monte Swann*

In this picture, the electrode was purposely held too far away from the base metal. Notice the arc light is brighter. Because the arc has spread out, there is a lot of spatter falling on the base metal. *Monte Swann*

Avoid jamming the rod down into the weld. If you are too close, hardly any arc light is visible. Short arc length will not adequately heat the base metal. *Monte Swann*

axis (side to side). It is very common for beginning and intermediate welders to change their electrode angle during the weld, especially when the bead is 6 inches or longer. No matter how short the electrode gets, pay attention to maintaining the correct angle during the entire length of the weld. The correct angle will change depending on the joint design and welding position. See each exercise for the recommended angles.

Relationship to *heat:*

The penetration into base metal will change depending on your electrode angle. Think of a garden hose spraying water in a muddy backyard. Holding the hose at a shallow angle to the ground will make a small dent, but most of the water bounces off in a spray. If the hose is directed at a steep angle, the water will penetrate much further into the ground. An SMAW electrode is similar, except you are spraying metal instead of water. The heat of the welding arc and the filler metal in the arc stream is directed by the angle of the electrode. Angles used in SMAW will determine both the amount of penetration into the base metal and where the heat is placed in the joint. When incorrect electrode angles are used, the heat will be directed to one part of the joint and not the other, making a much weaker weld.

Because of an incorrect rod angle, most of the heat is bouncing off the plate. *Monte Swann*

Steep travel angles will push your weld bead backward, resulting in limited penetration. *Monte Swann*

On this fillet weld in the flat position, several things are going wrong at once. The work angle is nowhere near 45 degrees and the amperage has been set too low. *Monte Swann*

Incorrect work angles cause undercut and a lack of fusion on one side of the weld bead. *Monte Swann*

The resulting weld has many discontinuities. There is lack of fusion and slag has filled in the voids. It is more common to see this at the beginning of a fillet weld because of quick starts and at the end because of an incorrect rod angle or arc length. *Monte Swann*

Speed

The key to creating uniform weld beads is a consistent travel speed. Any small variation in speed, like skipping ahead or a rapid movement side to side, will show in the final weld. Traveling too fast will result in shallow penetration into the base metal and a lack of adequate fusion. At a high enough amperage, a weld bead made too quickly can have a good appearance, but will lack strength when tested.

Relationship to **heat:**

Faster travel speeds will put less heat into the base metal and deposit less filler metal, making the bead smaller. Slower travel speeds will put more heat into the base metal and build up weld bead.

Heat

The volume of heat will depend upon the amount of amperage being used. Even though electrodes work over a wide range of amperages, welding at the upper or lower limits of the working range may cause problems in the final weld. A better way to increase or decrease the amount of heat required for welding is by switching to a larger- or smaller-diameter electrode. To determine overall heat input, look at the ripple pattern in the finished bead. Rounded, crescent-shaped ripples indicate the proper amount of heat. Pointy ripples are caused by too much heat when the center of the weld takes a longer time to cool. A flat weld profile is also too hot, while a tall weld bead is too cold.

No matter what position you are welding in, try to limit the plate temperature to 400 degrees F or less. There are many ways to measure plate temperatures, but a temp-stick is a cheap way to monitor this.

ARC BLOW

There is a strong magnetism created by welding with DC. As the magnetic field passes through the base metal it can become distorted. Arc blow happens when the arc is deflected from the weld area by this magnetic distortion. This is obvious when it happens because the welding arc wanders in a different direction, carrying the arc stream and filler metal with it. It is more common to have arc blow when welding in corners or at the end of a weld. To help prevent arc blow:

Amperage that is set too high boils the molten pool and creates a large amount of welding fumes. *Monte Swann*

This finished bead made with too-high amperage is elongated. Half the length is glowing orange from an excessive amount of heat. *Monte Swann*

Amperage set too low results in a tall, narrow bead because the filler metal is not flowing into the base metal. *Monte Swann*

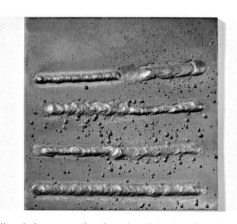

From bottom to top: proper bead, arc length too long/too short, incorrect travel and work angles, amperage too high/too low. *Monte Swann*

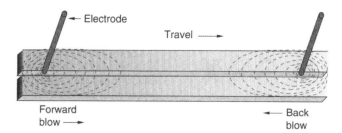

Electrode

Travel →

Forward blow →

← Back blow

Welding Inspection Technology, Fourth Edition, Figure 3.9, Reproduced with permission from the American Welding Society (AWS), Miami, FL USA

1. Lower the amperage—extreme heat from excessive amperages is a common cause of arc blow.
2. Position the work clamp at the end of a weld and close to the area of welding.
3. Use short arc lengths, slightly less than the diameter of the electrode.
4. Weld toward intermittent heavy tacks or to the end of another weld bead.
5. Use a back-step welding technique. See the square groove weld exercise in the GTAW chapter.
6. Completely eliminate arc blow by using AC instead of DC.

SETUP

With any project using SMAW, take a few minutes to get set up properly. Use some scrap metal of the same thickness as your project to experiment with the amperage setting. Don't be afraid to make changes to both amperage and arc control if your machine has those features. Have a variety of electrode sizes available so you can make adjustments to the amount of amperage (heat) used. Many different types of electrodes are available to suit almost any base metal and joint design. Take some time to select the correct electrode for your project and keep extras for future use. All the exercises in this section use DCEP. There is no harm in trying DCEN or AC with an electrode to see the difference in how it runs with those types of current.

Welding skills are immaterial if you are not using the electrodes and equipment correctly. Asking someone to weld together 0.030-inch thick metal with a ⅛-inch-diameter electrode is like asking a musician to play the piano with a broken thumb. Either way it will be difficult or impossible to do well. The most important thing is to relax and get comfortable. Remember ABC—always be comfortable. Rest your arm on the table, brace your body against a solid object, keep your elbows tucked into your sides, and sit down when possible. Wear the proper protective equipment so you're not distracted by the sparks hitting your gloves and arms or legs when sitting. When you are familiar and comfortable with the process, your chances of making great welds improve dramatically.

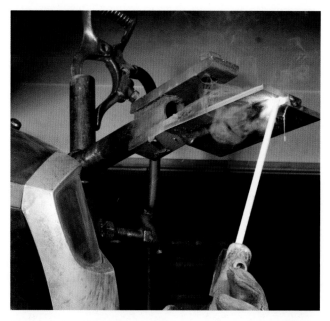

On this overhead lap joint, I traveled in a direction moving away from the work clamp attached at the top of the pole. Because of this, my arc was more erratic in the last half of the weld. *Monte Swann*

WELDING OUT OF POSITION
Vertical

Unless you are welding sheet metals ⅛-inch thick or less, always travel uphill with SMAW. Welding uphill will result in more penetration into the base metal. Cover passes on multi-pass welds can be made downhill, but this is not allowed in structural welding applications. Welding vertically up will require less heat. The weld pool will be more sensitive to the overall heat input. Use amperages on the low side of the working range along with short arc lengths. If the puddle drips out and molten metal runs down the joint, your amperage is too high.

On vertical and horizontal weld joints, the weld metal may appear to drip down. Don't let this fool you. The slag will droop and fall out of the joint when welding in these positions. This is normal, especially when using an E7018 electrode.

Overhead

For welding joints in the overhead position, use the same or slightly less amperage than for welding in the flat position. It works about the same way as welding in the flat position, only it is more difficult to maintain the proper arc length. Your arms will get tired quickly because you are no longer working with gravity, but are fighting it instead.

Be certain to dress properly. Protect your arms, chest, and the top of your head from sparks and hot slag. Slag can drip out of the weld and will burn anything it falls on, including your helmet. Wear ear plugs or some other type of ear protection to avoid the painful experience of having a spark or hot slag find its way into your ear canal.

When welding overhead, get as comfortable as possible. Keep your elbows in tight to your body and lean against a wall or table to steady yourself. *Monte Swann*

Drag the electrode toward the ground clamp to help keep the arc more stable while welding. *Monte Swann*

The slag will drip down. Don't let this mislead you into thinking that the weld metal is dripping out. *Monte Swann*

Keep a tight arc length. If the finished weld bead looks wavy with several high spots, it's because the arc length was not consistently maintained during the weld. *Monte Swann*

This is the finished fillet weld as it appears from underneath. *Monte Swann*

WELDING EXERCISES

I recommend completing these exercises in the order presented. The level of difficulty increases with each exercise, so become proficient with your new skills in one exercise before moving to the next. Wear the proper PPE (see chapter 2 for more information).

For practice metal, I recommend buying ¼ × 2-inch flat bar and use a torch to cut pieces 6 inches long. Or you can buy ¼ × 6-inch flat bar and use a saw to cut 2-inch strips or have them sheared to size. Either way, finished practice welds can be torch-cut apart, flipped around, and reused.

Since E6013 is an all-position electrode, all of the exercises except number 2 can be done in the horizontal, vertical, and overhead positions. Master each of the welds in the flat position before moving on to other positions.

Before beginning these exercises you will need:

- Constant current SMAW power supply
- ⅛-inch diameter E6013 electrodes
- ¼-inch mild-steel pieces cut approximately 2 × 6 inches
- Chipping hammer and wire brush for cleaning off slag
- Welding helmet with a No. 10 shade lens
- Pliers for handling hot metal
- Metal quench bucket filled with water to cool and inspect your finished work faster

EXERCISE 1
Bead on a plate

Electrode angles:

Work angle: 90 degrees

Travel angle: 60–75 degrees

Learning to run a bead on a plate is the first, most important exercise to master. Look for a bead that is uniform in width and height, with good fusion at the toes. Whenever you switch to a different size or type of electrode, running a bead on a plate will help you observe the welding action and set the proper amperage. When making a bead on a plate with E6013 you often get an inch of weld metal for an inch of melted electrode. Measure your bead and electrode to find out if your distance, angle, speed, and heat are correct.

Get comfortable before you begin. Use both hands to hold the electrode holder and rest one arm on the table. The electrode has been placed at a 45-degree angle in the grooves of the holder, making it easier to weld. *Monte Swann*

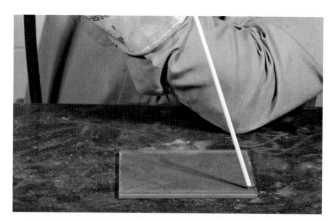

If you don't have an auto-darkening helmet, place the electrode as close to the beginning as possible before flipping down your helmet. *Monte Swann*

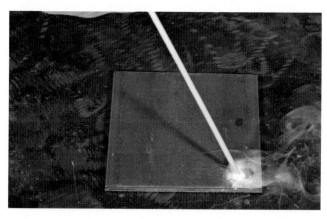

Strike an arc and circle the electrode once or twice to widen the bead and heat up the base metal. This plate heats up quickly because the weld was started in the corner. *Monte Swann*

Begin pulling the electrode along the plate. Maintain a close arc length and watch the formation of the weld pool. *Monte Swann*

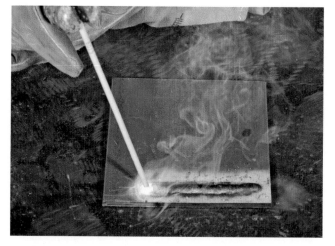

Use the edge of the metal as a guide to keep the weld bead straight. Try not to change your electrode angle throughout the weld. *Monte Swann*

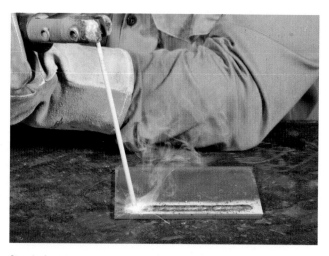

Stop before you reach the edge of the plate. Circle the electrode once or twice to add filler metal to the crater at the end of the weld. *Monte Swann*

EXERCISE 2
Layering weld beads
Electrode angles:

> Work angle: 80 degrees
>
> Travel angle: 60–75 degrees

The purpose of this exercise is to learn proper bead placement and create welds without slag inclusions.

Lay out a 5 × 5-inch square on a 7 × 7-inch piece of ⅜-inch mild steel. Begin the first layer by traveling right to left. After making a first pass, point the electrode at the toe of the previous pass making a second weld bead that is half on the base metal and half on the previous pass. Make weld beads half on and half off until enough rows are deposited to fill the 5 × 5-inch square. Be sure to chip and brush the slag off each pass. Quench the metal after every two or three passes to prevent overheating the base metal. Four other layers can be added on top. Orient each new layer perpendicular to the previous one to maintain a uniform height. Practice traveling from top to bottom, bottom to top, and left to right. Torch cut or saw cut the finished pad. Sand the edge smooth and inspect it for any slag inclusions or lack of fusion.

The sixth row of layered beads is being finished on this ⅜ × 7 × 7-inch piece of steel. *Monte Swann*

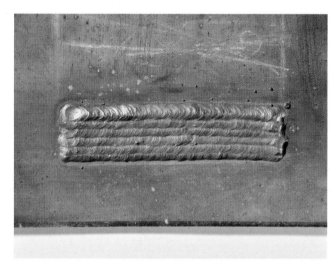

Layered weld beads should be made in a tight formation without gaps or slag inclusions between the passes. In the last pass, the weld bead was stopped and restarted about three-quarters of the way down. *Monte Swann*

Torch-cut out a center cube 2 × 2 inches, sand the sides smooth, and look for discontinuities. Notice the extreme amount of distortion on the ⅜-inch base plate. The cumulative shrinkage stresses of all the weld beads made on one side distorted the steel plate. *Monte Swann*

A band saw will also work for cutting the metal. Look for any voids, slag inclusions, or porosity. This weld shows complete fusion has taken place between all layers, except for in one spot in the lower-left corner. *Monte Swann*

EXERCISE 3
Lap joint, flat position
Electrode angles:

> Work angle: 45–55 degrees
> Travel angle: 60–75 degrees

Be certain you position the lap joint at a 45-degree angle from the table. With your hand in a dry glove, you will be able to hold the base metal while tacking two pieces together. Partially used 6010/6011 electrodes work great for tacking. Strike an arc and burn a small bead into both pieces, and pull away quickly. Hold onto the pieces you are tacking for a moment, allowing time for the tack to solidify. More amperage is required for a lap weld than a bead on a plate. Look for a finished weld bead with equal legs and no lack of fusion or slag inclusions. In lap joints, it is a best practice to melt the top edge completely or leave the top edge intact with the bead placed below it. As a guideline, with metals less than ¼-inch thick, the top edge should be melted; if the base metal is ¼-inch thick or more, leave ¹⁄₁₆ inch below the top edge. To prevent incomplete fusion and slag inclusions, don't let the molten slag get ahead of your weld puddle.

Like making a bead on a plate, keep both hands on the stinger and get comfortable. The lap joint is tacked together first, then tacked to a base plate at a 45-degree angle for a flat-position weld. *Monte Swann*

After striking the arc and starting the bead, begin pulling the electrode. Look at the top edge of the weld pool; it should look C-shaped at the end of the electrode. When you see this happening, it usually means the arc is penetrating into the root of the joint for proper fusion. *Monte Swann*

Maintain a consistent travel speed to fill in the joint. I've chosen to weld right to left on the joint, because it is more comfortable and I can clearly see the arc. *Monte Swann*

At the end of the weld, circle the rod a few times and keep a tight arc length to prevent blowing it out, leaving undercut or a concave crater. *Monte Swann*

Look for good fusion into both plates. Since this joint is filed up to the edge of the ¼-inch-thick plate, the bottom leg should be at least ¼ inch in size. *Monte Swann*

EXERCISE 4
Tee joint, flat position

Electrode angles:

Work angle: 45 degrees from each plate

Travel angle: 60–75 degrees

A tee joint requires slightly more amperage than the lap joint. Your finished weld should have equal legs running the entire length of the joint. Better results might be obtained by pulling the electrode. This will give you a better view of the molten puddle and help keep the slag away from the leading edge of your arc. The heavy slag coating makes the weld pool difficult to see. Look for the crescent shape at the back edge of the molten puddle. This indicates that the arc stream is penetrating into the base metals and the filler metal is flowing into the joint. Practice both the tee and lap joints at the same time.

I may look a little uncomfortable at the start of this weld, but as the electrode burns, I will have to lower my arms. Even with one elbow on the table, my forearms can become a pivot point, which will only allow me to keep the proper work and travel angles for part of the weld. *Monte Swann*

Don't tilt your electrode to one side. If your work angle is not 45 degrees from each plate, the weld will have unequal legs. *Monte Swann*

Because the fusion faces are in close proximity to each other and the volume of base metal will dissipate the heat, I've increased my amperage 20 percent higher than what I used for a bead on a plate. *Monte Swann*

Pulling the electrode will help keep the arc on the leading edge of the weld pool. Never let the molten slag get in front of your arc. *Monte Swann*

Notice the blue temper color on the sheared edge of the plate has followed along with the heat of the weld bead. *Monte Swann*

The slag may be more difficult to chip out of a tee joint. The weld should be uniform and have equal legs without slag inclusions or lack of fusion at the beginning or end. *Monte Swann*

EXERCISE 5
Open-corner flat-position with complete joint penetration (CJP)

Electrode angles:

Work angle: 90 degrees

Travel angle: 60–75 degrees

The amperage for this exercise is the same as exercise number 1. In the first pass, also called the root pass, the electrode must be pulled. Chip, brush, and let the first pass air cool or quench in water to save time. The second pass, also called the cover pass, is made with a slight weaving motion to join both edges along the corner. The most important consideration with this exercise is the relationship between the amount of root opening and amperage setting. A smaller root opening (1/16 inch or less) will be able to carry more heat and higher amperage. A larger root opening will require less amperage.

Hold the stinger loosely in your hands. A tight grip is not necessary to make a good weld. *Monte Swann*

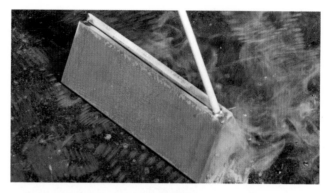

At the beginning of the weld, push the burning electrode through the root opening of the joint. This should open up a small keyhole between the two pieces. *Monte Swann*

Once you are all the way through, bring the rod up a little and begin pulling. The weld bead should fill the opening as you go. *Monte Swann*

If the hole becomes too large to fill, stop. It may be necessary to turn your amperage down or reduce the amount of root opening, in which case you'll need to tack a different corner together. *Monte Swann*

It is unnecessary to circle the electrode after the first pass. Move the electrode to the end of the joint and pull out. *Monte Swann*

You should be able to chip the slag from the other side of the joint. A small bead of filler metal here indicates CJP. *Monte Swann*

To have a better view of the weld pool on this tee joint, position your head in front of the joint and prop the weld up off the table. *Monte Swann*

After finishing the first pass, chip and brush the slag and allow the weld to cool. Make the second pass with a slight weaving motion to fill in the top, tying in the edges. *Monte Swann*

I knew there would be trouble when I traveled too fast at the beginning and didn't see that important C-shaped weld pool right away. *Monte Swann*

Everything is working OK in the middle, which is typical when making a 6-inch-long weld bead. *Monte Swann*

EXERCISE 6

Tee joint, horizontal position

Electrode angles:

Work angle: 35–40 degrees from the bottom plate

Travel angle: 60–75 degrees

The purpose of this exercise is to learn to compensate for the effects of gravity on the molten weld pool. A tee joint welded in the horizontal position is difficult because it is easy to undercut the top plate using an incorrect work angle. To correct the work angle, point the electrode toward the side with the undercut. Another common problem is ending up with too much molten filler metal deposited on the base plate (unequal legs).

At the end of the weld, it looks like I've changed my work angle from 35 degrees to 45 degrees. Since I'm no longer compensating for gravity, part of the finished weld will have unequal legs. *Monte Swann*

The end of this weld glows orange because the filler metal is solidifying and cooling underneath the slag covering, protected from atmospheric contamination. *Monte Swann*

Notice the bad start to the weld. The very beginning shows where fusion did not take place at the root of the joint and filled in with slag. In order to fix it, grind out the bad start and reweld that area. The small, circular dents in the face of the fillet weld were made by the sharp point of the chipping hammer as the slag was removed. *Monte Swann*

The stacked-dime appearance of this E-6011 bead and fillet weld is made by using the whip technique. *Monte Swann*

WELDING WITH E6010/E6011

These types of electrodes are also called whip rods. They are fast-freeze electrodes, meaning the filler metal solidifies very quickly. This allows you to use different welding techniques. The whipping technique involves depositing weld metal in sections instead of in one continuous motion like with E6013. Since these electrodes have a cellulose-based flux, less current is needed for welding. Use less amperage with a 6010/6011 electrode than for 6013 or 7018. Strike an arc and start the bead by circling once or twice. Move the electrode just out of the puddle, allowing the filler metal to solidify. Move the end of the electrode back, slightly ahead of where you started. This is all done in quick motions. Keep moving away from and back to the weld pool as you travel down the joint. The finished bead should have the appearance of stacked dimes.

On a tee joint with E6010 or E6011, when you bring the electrode back to the weld pool to deposit more metal, wait a moment. The longer you stay over one spot, the more filler metal will flow into the joint and build up weld bead. Remember to keep a tight arc length as you deposit filler metal.

Practice beads, lap, tee, and open corner joints in all positions.

Exercise with E6011-single V-groove, open root
⅜-inch thick mild-steel plate cut 3 × 6 inch

Bevel each plate 30 degrees (60-degree groove angle), with a 1/16-inch root face

Root opening 1/16 inch

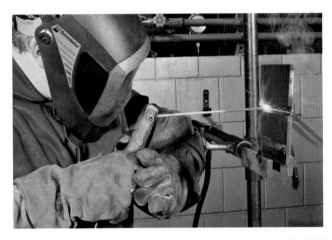

As with the open corner joint, begin this V-groove by punching the 6011 electrode through the root of the joint. *Monte Swann*

Use a C-shaped motion instead of a whip to reduce the possibility of slag inclusions. As you move the electrode, watch the root of the joint fill in. If the opening becomes too large or too fluid, try decreasing the amperage or using a smaller root opening. Don't forget to sand a flat spot (root face) on each beveled edge to carry the heat. *Monte Swann*

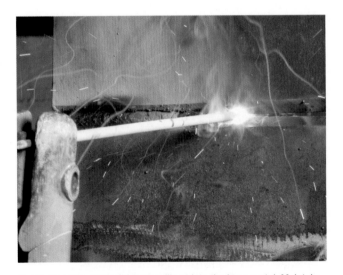

Remember: The heat of the arc will cut into the base metal. Maintain a tight arc and consistent travel speed despite your instinct to pull the rod away. *Monte Swann*

The root opening may begin to close at the end. Leave a slightly larger gap at one end to compensate for this. *Monte Swann*

Filling in the end of the joint was difficult. It could be fixed by chipping and brushing the slag, grinding out a short transition to the weld bead, and rewelding. *Monte Swann*

This root pass has one restart in the center, which is a little rough and below flush on the back side. *Monte Swann*

After making the root pass, this joint can be finished by layering filler and cover passes in a sequence from bottom to top. *Monte Swann*

WELDING WITH E7018

Any electrode with the last digit of 5, 6, or 8 is in the low-hydrogen group. These electrodes are especially designed to limit the amount of hydrogen in the finished weld. For more information on why this is important, read the section on HAZ cracks in the chapter on discontinuities. In order for E7018 to prevent hydrogen from being deposited in the weld bead, the flux on the electrode must be kept dry. Once the container is opened, the flux will begin to absorb moisture out of the air. It takes about four hours of exposure to the air before the properties of the flux coating are compromised. If this happens, the electrode will no longer work as well as it should, especially on high-strength and low-alloy steels. If you are using an electrode with the new H4R suffix, the electrode is good for nine hours instead of four.

A drive rod oven, like this one, is designed to store 7018 low-hydrogen electrodes. *Monte Swann*

For this reason, 7018 electrodes are kept in a dry-rod oven, which bakes the electrodes at 250 degrees F. Electrodes left out in the air can be rebaked once at 700 degrees for one hour to restore the properties of the flux. It is possible to weld mild steel without baking E7018, but the finished weld will not be as strong.

EXERCISE WITH E7018
Single V-groove with backing
⅜-inch-thick mild-steel plate cut 3½ × 7 inches

    Backup bar ¼ × 1½-inch flat bar, cut 9 inches long (sand mill scale off of one side)

    Bevel each plate 22.5–25 degrees (45–55 degree groove angle)

    Root opening ¼ inch

Start the arc on the runoff tab. Use lower amperages for vertical up welding. Keep a tight arc length and a travel angle of 85 to 90 degrees. *Monte Swann*

Move the electrode back and forth, centering it over the beveled edges. As an alternative, a single pass can be made over each beveled edge with a 3/32-inch-diameter electrode. *Monte Swann*

Since I'm using a slight weave motion for the root pass to weld together both plates and the backing bar, I'll run out of electrode before reaching the top. I will finish the rest of the bead using a second electrode. *Monte Swann*

CHAPTER 11
GAS METAL ARC WELDING (GMAW)

The term "wirefeed" is used to refer to a variety of welding processes. Wirefeed describes how the electrode is delivered to the weld pool. The welding wire (electrode) is unwound from a spool and fed automatically through the gun cable to the molten pool during welding. This type of eletrode is a continuous consumable electrode. Though all wirefeed processes have this continuous consumable electrode in common, that is where the similarities end. There are many variations within the umbrella of wirefeed welding. In this chapter, we will look at several wirefeed processes, concentrating on GMAW short-circuit transfer.

Metal inert gas (MIG) is a commonly used term instead of gas metal arc welding (GMAW). But carbon dioxide is used as a shielding gas in some types of wirefeed welding. By definition, carbon dioxide is not an inert gas (like argon) because it is comprised of more than one type of atom. But it acts like an inert gas by not combining with other elements in the atmosphere. Since carbon dioxide is not a true inert gas, MIG is a less accurate way to describe the process. Chart 11-1 shows a wirefeed "family tree," illustrating the relationships between the various wirefeed processes.

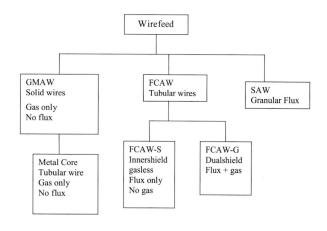

GMAW - Gas metal arc welding
FCAW - Flux cored arc welding
SAW - Submerged arc welding

Chart 11-1

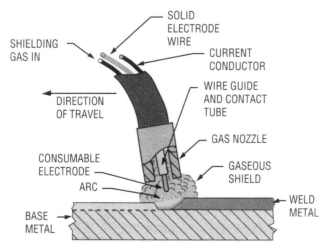

Welding Inspection Technology, Fourth Edition, Figure 3.10, Reproduced with permission from the American Welding Society (AWS), Miami, FL USA

An old-school welding instructor, nicknamed "Sergeant Slag," once said, "Wirefeed was the worst thing that ever happened to welding." He had seen the development of wirefeed in a time before I was born. I asked him why he thought this way. His reason: because wirefeed made welding a lot easier than before and allowed less-skilled workers to perform it. I can understand his perspective. I've been at dinner parties where, upon finding out my profession, someone informs me that their five-year-old is able to weld, too. I believe a five-year-old could weld using a wirefeed process. It is that easy. But knowing how to set up the machine; select the proper electrode and shielding gas; dial in the settings for the type of joint, metal thickness, and position of welding; and evaluating the final weld for discontinuities takes a little more skill.

Wirefeed was a blessing to industry. Faster travel speeds and higher deposition rates were possible, making joining metals cheaper and more efficient. Robotic welding relies on wirefeed as the process that best suits automated welding systems. Sergeant Slag is right—wirefeed lowered the bar in the level of welding skill required for success. But other processes, still in wide use today, have certain advantages over wirefeed welding. Is wirefeed the right process for you? It may be, but choose the process based upon what you plan to do with it. Choose wirefeed because it best suits your needs and not because it is easiest to learn.

GMAW

Gas metal arc welding (GMAW) is one type of wirefeed welding. Each term in the name refers to:

Gas

A shielding gas is used to protect the molten weld pool from atmospheric contamination. The wire is solid, there is no flux involved with the process, and no slag byproduct to deal with. A wide variety of shielding gases are used in both GMAW and FCAW-G.

Metal

Almost any metal can be welded with GMAW, including, steel, stainless steel, aluminum, and copper alloys.

Arc

Like all arc-welding processes, electricity is used to generate the heat required for welding. There are a variety of welding arcs used in GMAW. These different arc characteristics are called modes of metal transfer.

Welding

GMAW is a semiautomatic process where the filler metal is fed automatically into the weld pool. Both the welding wire (electrode) and base metal must be hot enough for proper fusion to take place.

GMAW/MIG MACHINES

A CV power source will be required for any of the wirefeed welding processes. CC machines can be adapted to provide CV, but this is not an accepted practice in structural steel welding. See chapter 9 for more information on CV machines.

MACHINE CONTROLS

Polarity

For GMAW (welding with solid wires) DCEP is always used. That means the gun cable (welding lead) is hooked up to the positive terminal and the work clamp is hooked to the negative terminal on the machine. Using DCEN for GMAW will result in a wide shallow bead with little penetration into the base metal and excessive amounts of spatter. DCEN is used with some tubular wires in FCAW, depending on the chemical composition of the flux in the electrode. AC is not used in GMAW or FCAW.

Voltage

A CV power source is used for wirefeed welding. Unlike CC machines (used for SMAW and GTAW) CV machines have a control for setting the voltage amount. Although voltage is the force behind amperage, it is useful to think of it functioning like amperage, controlling the amount of heat. The arc length is self-regulating in a CV power source, meaning that the machine will make adjustments automatically for variations in arc length (distance).

Wirefeed speed and gas flow

With a CV power source, the amount of amperage is determined by the wirefeed speed (WFS). In other words, the wirefeed speed control is the amperage control. WFS is the rate at which wire is being fed to the weld pool. Although amperage is the amount of current flowing in the circuit, it is useful to think of wirefeed speed as an adjustment made secondary to voltage. WFS is measured in inches of wire fed per minute (IPM). Machines with an inch button will feed wire without gas flow or electrical output.

Shielding gas-flow rates are measured in cubic feet per hour (CFH). Flow rates are set on the regulator/flow meter. Machines with a purge button can make gas flow without the wire feeding or electrical output. Machines without either inch or purge controls will need to have the gun trigger depressed to make gas flow or wire feed. To feed wire with this type of machine, turn off the gas cylinder or flow meter to prevent wasting gas. To adjust gas flow, release the tension adjustment screw so wire is not fed through the drive rolls.

In either case, remember the wire will be electrically charged and will arc on anything connected to the welding circuit. I have made the mistake of setting the gun down so the trigger rested on the table, inadvertently feeding wire out while making an adjustment on the machine. The wire came in contact with a conductable work surface, fused itself to the metal, and heated very quickly. It was no fun grabbing that wire to clear it away, and discovering that you can be burned—even when you're not welding.

Inductance

Some older machines have an inductance control. Inductance control changes how fast the current rises in short-circuit transfer welding. No matter if the current rises slowly or quickly, there is no change to the final amount of current available. Increased inductance supplies energy to the arc after the short has cleared, increasing the time the arc lasts. Increasing inductance makes the puddle more fluid and flatter, producing smoother weld beads. Most new welding machines control inductance automatically when welding with short-circuit transfer.

Slope control

When the relationship between voltage and amperage is plotted on a curve, it is generally referred to as volt change per 100 amps. The shape of this curve determines the characteristics of the welding arc. A particular type of slope (curve) is provided by a power supply, depending on the welding process it is used for. A modified volt/amp curve with a steeper slope is used in short-circuit transfer. The volt/amp curve can also be flattened for use with globular and spray transfers (see chapter 9 for more information on the v/a curve). On a machine with slope control, the volt/amp curve can be modified to provide more or less amperage for a given amount of voltage. For example, if a machine is set for a certain voltage, a steeper slope provides less amperage and less heat to the welding arc. A flattened slope provides more amperage and more heat for the same voltage setting. Most new welding machines will automatically adjust slope depending on where the other machine controls are set and the type of wirefeed welding being done.

EQUIPMENT

More equipment is required for wirefeed welding than for any other manual-welding process. Before you purchase a machine, look into all the accessories available for it.

Guns

The torch used in wirefeed welding is commonly referred to as a gun. Like the torch used in gas welding and GTAW, the gun is the business end of the welding machine. The gun carries the welding current, shielding gas, and wire to the weld pool. When you depress the trigger on the gun, it sends a signal back to the machine to begin all three of these actions at once. Keep your gun in good condition and have replacement parts and consumables on hand so small problems don't become major setbacks.

Several brands of wirefeed guns are available. A gun will be supplied with the machine, or you can purchase brand-name guns (Tweeco is one of the better-known brands), which may work better for your applications. These guns are fitted with different ends that attach to the various brand-name welding machines such as Miller, Lincoln, and ESAB. Adaptors are also made so guns can work with a variety of machines.

Wirefeed guns are either air-cooled or water-cooled. Water-cooled guns are used in 200-plus-amp high-heat and high-deposition welding. These amperages are not typical for most welding projects done in a small shop. Air-cooled guns are the most common type, and they are lighter and less expensive than water-cooled guns. Guns are rated for the maximum amount of amps they can handle. For example, a Tweeco No. 2 gun is rated for 200 amps. Air-cooled guns not

The size and shape of the nozzle, insulator, diffuser, and contact tip will depend on the brand and size of wirefeed gun. Diffusers and nozzles are made of either brass or copper. The two dark pieces are insulators, used with the longer nozzles in a two-part system. The shorter brass and copper nozzles have the insulators built in. *Monte Swann*

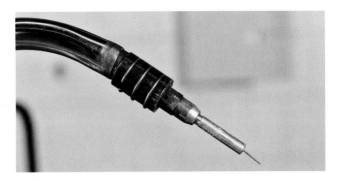

With the nozzle removed from the end of the gun, you can see how the insulator, diffuser, and contact tip fit together. *Monte Swann*

rated high enough for the job being done will get hot if large welds are deposited with high levels of heat. Buy a gun suited for the work you will be doing. It is better to use an overrated gun rather than underrated. The gun supplied with a machine should be the proper size for the maximum output voltage, but that gun may not always meet your welding requirements.

Nozzles and insulators

Nozzles direct the flow of shielding gas to the weld pool. Nozzles are made of copper or brass, and must be insulated from the welding current. Some nozzles have the insulator built in. Others are designed to be used with a separate insulator. The advantage of using the separate nozzle and insulator system is the adjustable contact tube setback. Nozzles with the insulator built in are limited in their adjustability. The position of the contact tip in relation to the nozzle impacts the amount of electrode extension and amount of heat going into the contact tip. Separate nozzle and insulator parts are available for guns that come with the one-piece system. Nozzles also come in different sizes; some are tapered at the end for greater access in difficult-to-reach places or in narrow joints.

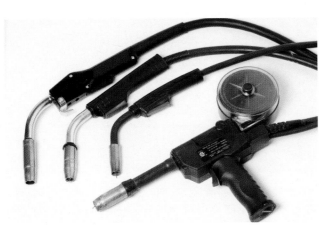

The three air-cooled guns on the left are different sizes, brands, and amperage ratings. Small spools of aluminum, stainless steel, and silicon bronze wire are used in the spool gun pictured at bottom right. *Monte Swann*

Spatter from welding collects on the contact tip and nozzle, eventually clogging the end and restricting the flow of shielding gas. Nozzle dip (welder's jelly) or anti-spatter spray is used to keep weld spatter from collecting on these gun parts for a longer period of time.

The end of this contact tip is worn out and full of spatter from welding. It's a good thing that I have plenty of new ones to replace it when needed. *Monte Swann*

When you replace the contact tip, make sure that it is the correct size for the wire diameter being used. The numbers are embossed on every tip. Use a pair of pliers to tighten it down. A loose contact tip can cause an erratic arc. *Monte Swann*

Weld spatter will build up on the end of a nozzle. Because the nozzle is made of copper, the spatter will come off easily. A small rat-tail file is a good tool for this job, but only if your nozzle and insulator are separate pieces. *Monte Swann*

Diffuser

The diffuser has small holes machined into the sides. The holes permit the flow of shielding gas from the gun cable to the nozzle. The holes should always be kept clear of debris and unblocked by other gun parts. Blocked diffuser holes will not allow for adequate shielding gas flow. Attached to the end of the diffuser is the contact tip (tube). The diffuser carries the current to the contact tube and is electrically charged while the arc is on. If an arc is struck accidentally on the diffuser, the soft brass part will melt and probably ruin the part. When changing gun consumables, it is a good idea to turn off the machine. A small hex screw on the diffuser secures the end of the gun liner to the diffuser. This screw needs to be loosened in order to change the diffuser or gun liner.

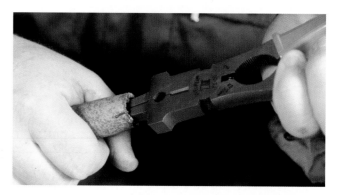

A pair of pliers works well for cleaning the nozzle. These welpers are specifically designed for working on wirefeed guns. Using a file on a nozzle with a built-in insulator may damage the threads. Pliers are the best alternative. *Monte Swann*

Nozzle dip or anti-spatter spray keeps spatter from collecting on gun parts. Before I dip the gun into this can of welding jelly, I run a couple of beads to heat up the parts. A cold nozzle and contact tip will pick up a big clump of jelly. If this happens, you can take off the nozzle and clean the gun parts with a towel, or burn it off by making welds on a piece of scrap. If you choose to burn it off, be prepared for a lot of fumes and mess. *Monte Swann*

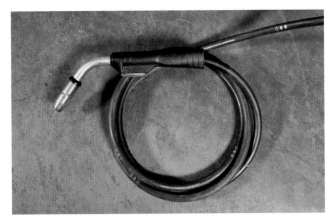

Do not tightly coil your gun cable. Storing it this way can damage the liner and wear it out faster. Never put a sharp bend or kink in your gun cable. Keep it as loose as possible when welding or storing it. *Monte Swann*

Contact tips (contact tube)

The most common consumable on a wirefeed gun is the contact tip, also called contact tube. These tips are designed to wear out and are replaced often. Contact tips need to be in good condition to maintain a stable arc. Contact tips are sized for the diameter of wire, so a 0.035-inch (9mm) contact tip is used for a 0.035-inch diameter electrode. In some cases, it may be useful to use a contact tip one size larger than the diameter of the welding wire. For example, when using aluminum wires or spray transfer, a large contact tip leaves room for wire expansion from the heat of the arc.

The hole at the end of the copper alloy contact tip gets worn, oval-shaped, and out of round. The tip can also become contaminated with spatter, even if nozzle dip is used to extend its life. Since the welding current is transferred from the contact tip to the wire, a dirty or worn tip can cause the arc to become erratic. Contact tips are inexpensive and should be replaced as needed.

Work (ground) clamp

The work clamp is rated for the amount of amperage used. The correct-sized work clamp for welding prevents resistance in the circuit and heat buildup. Always be certain your work clamp is firmly attached to bare metal and is as close to the welding zone as possible on larger pieces. Work clamps attached to dirty, rusty, or painted metals can cause the welding arc to become erratic.

Gun cable and liner

Inside the gun cable is a copper lead to carry the welding current, hose to deliver the shielding gas, and a steel coiled liner to carry the welding wire through the cable. Gun liners wear out over time and need to be replaced if kinked or plugged with debris. A gun cable coiled tightly or twisted and bent during welding will reduce the working life of the part. Use care not to pull liners around sharp corners or run them

over with a heavy vehicle. A bad liner will impact how well the wire is fed through to the work. Irregular wirefeed speed, an erratic welding arc, burnback, and birdnesting can all result from a dirty, kinked, or worn liner. If the gun or cables get hot during welding, check to make certain the cables are not too small or too long for the amperage being used. Upgrade to a higher rated gun when necessary.

Drive rolls

All wirefeed machines have at least one set of drive rolls. One or more of these rolls are attached to the wirefeed motor. The speed at which they turn is determined by the WFS set on the machine. A spring-loaded tension adjustment screw puts pressure on the top drive roll. This pinches the wire between the two rolls so that it can be fed off the spool and through the gun cable to the work.

Drive rolls, gun liners, and inlet guides are all important parts of a wirefeed system. The blue and white spool is 1 pound of 0.030-diameter 4043 aluminum welding wire. The small spool fits into a spool gun, where the gun and wirefeed drive-roll system are combined. *Monte Swann*

Felt pads like this one are clipped on the welding wire and soaked with lubricant. This reduces wear on the drive rolls. It also prevents dust and dirt on the wire from clogging the gun cable liner. *Monte Swann*

Grooves are machined in the drive rolls, so that the wire is kept in place during feeding. Some drive-roll configurations used to feed solid-steel wires will have one grooved roller on the bottom and a flat (non-grooved) roller on the top. Roller grooves can be V-shaped, U-shaped, and ribbed (knurled) for added grip on cored wires. The grooves are also sized according to the diameter of wire being used. For a 0.030-inch-diameter wire, a 0.030-inch drive roll is ideal. One size larger, a 0.035-inch drive roll, will also work. But one size smaller (0.023 inch) will not. The drive roll sizes are embossed in the rollers.

This inlet guide has been pulled from the drive rolls. Notice the O-rings and holes for shielding gas. If this end of the gun cable is loose, or not pushed in all the way, the gas will not be able to get from the solenoid in the welding machine to the gun cable. Check this connection and the O-rings as one possible reason for gas-flow restrictions and porosity in your weld. *Monte Swann*

With the drive-roll tension screw released, you can see how the wire runs through the inlet guides and over the drive roll. With the gun cable hooked to the positive terminal, and ground hooked to the negative, the polarity is DCEP. *Monte Swann*

This midsized Miller CV wirefeed welder has a single set of large-diameter drive rolls. Notice the lug connection for the positive terminal is color-coded red. *Monte Swann*

Most small machines will have a single set of drive rolls. Larger machines may come with dual drive rolls to help feed wire over long lengths of gun cable. Dual-drive-roll systems greatly reduce the incidence of birdnesting, where the wire wraps around the drive rolls. Most drive-roll systems are push type, where the wire is pushed through the gun cable. An alternative is the pull type, where wire is pulled through the welding cable by drive rolls mounted in the gun. Push-pull systems employ two sets of drive rolls, one at each end of the gun cable. The feeder motors are synergic, which means they work together to feed wire, so only one WFS is set on the machine for both drive rolls. The Cobramatic wirefeed gun is one example of a push-pull type system.

Spool guns are different from the other types. A small spool of wire is mounted on the gun and a small set of drive rolls feeds the wire out to the contact tip and welding area. Push-type systems are used for solid or tubular steel wires. Softer wires, such as aluminum, and difficult-to-feed wires, such as stainless steel, use an alternative system, such as push-pull or a spool gun.

Inlet/outlet guides

The wire is pulled through a set of guides mounted on either side of the drive rolls. These guides align the wire coming into and going out of the drive rolls. The ends of these guides should be positioned as close to the drive rolls as possible without touching them. Missing or misaligned inlet/outlet guides may cause birdnesting.

Regulators and flow meters

In GMAW and FCAW-G, a shielding gas is always used. There are several varieties of single and mixed gases for use with the various types of wires and processes. The regulator must be

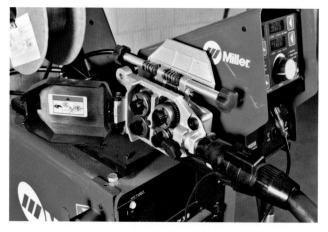

Large machines will have a dual-drive-roll system. Two sets of rollers, side by side, reduce wirefeed problems, such as birdnesting. *Monte Swann*

Shielding-gas regulators come in two styles. The gauge-type is similar to regulators used in oxy/acetylene, with a center adjustment screw used to set the gas-flow rate. The regulator/flow meter on the right is a glass-tube type. The knob on the side adjusts the flow of shielding gas. *Monte Swann*

The cylinder valve connections are different for mixed gas and straight CO_2 regulators. CO_2 regulators use a small nylon washer between the regulator and cylinder. If this washer is missing, CO_2 will quickly leak from the connection. *Monte Swann*

This glass-tube type regulator is used for CO_2 only. CO_2 regulators are specially designed to work at the low temperatures created when liquid CO_2 recompresses into a gas. *Monte Swann*

matched to the type of gas being used. Some regulators can be used with multiple types of gases, while others are designed for a specific type of gas, like CO_2. CO_2 regulators are designed not to freeze up from the cold temperatures associated with recompressing liquid carbon dioxide into a gas. Flow meters are typically combined with regulators, indicating the exact flow of shielding gas in cubic feet per hour (cfh).

Wirefeed transfers

In GMAW, the welding wire is transferred to the welding pool in a variety of ways. The method of transfer depends upon the voltage being used while the arc is on. Voltage set on the machine will drop when welding begins because of the electric load put on the machine. When the arc is off, the voltage is called open-circuit voltage. When the arc is on, it is called many things including: closed-circuit voltage, load voltage, actual voltage, working voltage, or arc voltage. All wirefeed transfers happen at a specific load voltage.

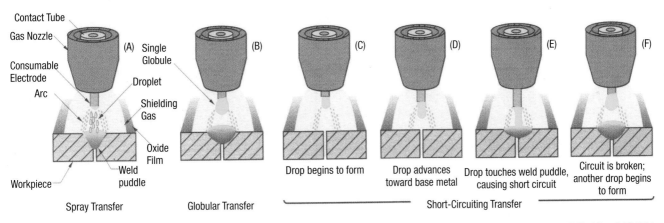

Welding Inspection Technology, Fourth Edition, Figure 3.14, Reproduced with permission from the American Welding Society (AWS), Miami, FL USA

- *Short-circuit transfer*
(15–22 load volts) also known as short arc

  In short-circuit transfer, the wire actually touches the metal. When the wire comes in contact with the metal, the electrical resistance causes the wire to heat up and pinch off. This heating of the wire is called inductance, which pinches off a small droplet of metal. The heat of the arc fuses it into the base metal. Since the machine is automatically feeding, the wire is fed back into contact with the metal. Another droplet of wire is pinched off and deposited. This process happens very quickly, about 20 to 200 times per second, depending on machine settings and amount of wire stickout. This is why short-circuit transfer makes a sizzling or bacon frying sound when the machine is dialed in correctly.

  Since the arc is not continually on during short-circuit transfer, the weld puddle is small and relatively cool. The heat input is lower than any of the other modes of transfer and the pool solidifies quickly. When the filler metal solidifies quickly it is described as a fast-freezing puddle. This makes short circuit useful for welding thinner gauge metals (³⁄₁₆ inch or less), filling joints with gaps or poor fit up, and welding in the vertical and overhead positions. Thicker metals can be welded with short-circuit transfer using proper joint design and a machine with a higher output current. It also requires high skill level to be done successfully.

- *Globular transfer*
(22–24 load volts)

  When voltages are increased from short circuit, droplets begin to form at the end of the wire. When these droplets become larger than the diameter of the electrode, it destabilizes the arc. Eventually, the large droplet either reconnects with the work piece or spins off the end of the wire and into the puddle, resulting in an erratic arc and lots of spatter. Globular transfer should be avoided with GMAW. However, in FCAW, the flux stabilizes the arc so proper transfer can take place. Droplets are pinched off the end of the flux-cored wire and transferred to the base metal through the continuous arc stream. With the proper electrode, FCAW-G or FCAW-S can be used in all positions on thicker sections of steel over ³⁄₁₆ inch in either the globular or spray voltage range.

- *Spray transfer*
(24–30 load volts)

  When the voltage increases past 24 and the proper shielding gas is used, the droplets become equal to or smaller than the diameter of the electrode. The finer droplets are sprayed across the arc in a cone-shaped stream and deposited into the base metal. When working properly, the continuous arc makes a humming sound. Spray transfer is used to deposit large amounts of filler metal on thicker materials over ⅛ inch. Spray transfer on steel creates a weld pool so fluid, it can only be used

for the flat and horizontal positions. On aluminum, the force of the arc stream can overcome gravity, making it possible to weld in all positions when the voltage and WFS are set correctly. Any machine with a high enough output range can perform short circuit, globular, and spray transfer.

- *Pulse-spray transfer*

  Pulse-spray transfer is a recent development in wirefeed welding. Not all CV power sources are able to perform pulse spray. The welding machine used for this mode of transfer will have specific controls. In pulse spray, the load voltage (and current) alternates between less than the short-circuit range (background) and spray range (peak). During peak current, a small droplet is melted off the end of the wire and propelled through the arc stream and into the weld pool. The metal transfer takes place only during peak current. During background current, no metal is transferred through the arc.

  One cycle of peak and background current is called a pulse. The amount of time that is spent at the peak and background currents can be changed, along with the output voltage (background current and amplitude). The number of the pulses in pulse spray are measured in pulses per second (PPS). On the machine, the PPS and WFS are set depending on the type/size of wire and the shielding gas being used.

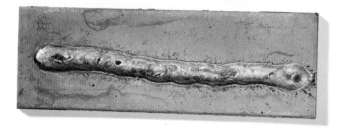

This is a GMAW weld with spray transfer. The arc is brighter and more intense because higher voltage and amperage settings are used for spray transfer. *Monte Swann*

In pulse spray, the arc is continuous, but the filler metal is deposited intermittently. This allows greater control of the amount of heat in the weld pool and base metal. Pulse spray has spray-like transfer in all welding positions, even when larger-diameter wires are used. A variety of metal thicknesses, from 20-gauge sheet to 1½-inch metals can be welded with a minimal amount of distortion.

ADVANTAGES

All wirefeed processes have certain advantages. Since the electrode is continually fed into the weld pool, the bead can be made without stopping. Fewer stops and starts reduce the chances of weld discontinuities, which usually occur at the beginning or end of a weld. The travel speeds are faster with wirefeed, so there is less heat input to the base metal. Generally the HAZ is smaller with GMAW than with gas welding, SMAW, or GTAW. The process is also easier to learn.

DISADVANTAGES

Although a wide variety of metals can be welded with GMAW/FCAW, the setup for each type of metal is different. Drive-roll systems, gun parts, welding cables, shielding gases, and regulators change depending on if you want to weld aluminum, stainless steel, copper alloys, nickel alloys, or mild steel. A spool gun is the easiest way to fulfill the drive-roll and feeding requirements of the different types of wire. The small spool of wire used with a spool gun reduces the amount of investment in filler metal. Larger spools cost less per pound, but require more costly equipment to run them properly.

Since a shielding gas is used with GMAW and FCAW-G, these processes can be disrupted by a breeze and are mostly used indoors. FCAW-S does not require a shielding gas. With a high output power source and large-diameter wire, FCAW-S is comparable to SMAW. Too often, a small-diameter electrode is used with low output power sources. This type of gasless wirefeed welding is a less effective way of joining metals compared to SMAW.

Wirefeed systems have many components. A power source, wirefeed motor, drive rolls, gas solenoid, gas regulator, welding leads, and gun are involved. More parts are used when welding with GMAW, and more things can malfunction or break.

ELECTRODES

The single biggest factor in selecting a welding electrode is to match the composition and mechanical requirements of the base metal. Other factors include: the service conditions, such as temperature and impact or abrasion resistance; cleanliness of the base metal; the position of welding; and which mode of transfer is going to be used. Find out which wire is the best suited for your particular project.

Wirefeed is broken down into categories according to wire type. Solid wires are used in GMAW. The most popular solid wire for welding steel is ER 70S-6.

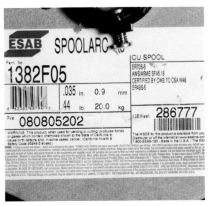

The manufacturer of a wire electrode assigns a product name for each type of electrode. Spoolarc 86 is the ESAB version of an ER70S-6 wire. Other information on the label includes the wire diameter, heat number, and spool size given in weight. *Monte Swann*

- **ER** stands for electrode/rod, meaning that the same number designation is used for either a spool of wire (electrode in wirefeed) or filler rod (used in GTAW or gas welding)
- **70** = 70,000 psi tensile strength
- **S** = solid wire
- **6** = type and amount of cleansing and deoxidizing agents in the electrode

See the chapter on GTAW for more information on the various solid wires and the meaning of their AWS number designations.

The number designation for any wire has to do with the characteristics of the wire rather than the wire's size. Typical wire sizes are 0.023-inch, 0.030-inch, 0.035-inch, and 0.045-inch diameter. There are larger wire sizes available for heavy industrial use. Only a certain number of electrons can travel through the welding wire and arc. As the diameter of the wire increases, its area also increases and a greater number of electrons can flow across. A 0.045-inch-diameter wire will allow more electrons to flow across than a 0.035-inch wire. Therefore, a smaller-diameter electrode will have less amperage capacity and heat. Thicker metals, requiring more heat for proper fusion to take place, should be welded with larger-diameter electrodes. The amperage output capacity of the machine determines what size wires can be used and what thickness of metals can be welded.

Tubular wires are used in FCAW. To make the wire, a flat metal strip is formed into a U-shape. Then a powdered flux is poured in and the wire is closed up, giving it a flux core. The ingredients and chemical composition of the flux added into the wire can be changed. Wires used for FCAW-G cost half as much as those used for FCAW-S, but do require a shielding gas in order to work properly. A typical FCAW wire is E71T-1 (AWS designation).

- **E** = electrode
- **71** = 71,000 psi tensile strength of the filler metal
- **T** = tubular wire
- **1** = chemical composition of the flux

SHIELDING GASES

Compressed-gas cylinders are potentially very dangerous and should be handled with care. Read the chapter on compressed gases for more information on cylinder safety. The type of shielding gas used in GMAW or FCAW-G will have a great effect on the depth of penetration (fusion) into the base metal. Shielding gases can also affect the characteristics of the welding arc and mechanical properties of the finished weld. The single most important rule in selecting the type of shielding gas is to match the gas to your wire. The wire manufacturer will give recommendations on which type of shielding gas to use. They may give several options, ranking the choices from best to least best. If you don't have the manufacturer's recommendations, ask your welding supplier for the information on recommended shielding gases for the type of wire you are using.

GMAW short-circuit transfer with a mild-steel wire

Straight CO_2 is least expensive and will penetrate further into the base metal. Welding with CO_2 produces more smoke and fumes than C25. The finished weld will have a rougher appearance.

C25 (75 percent argon and 25 percent CO_2) is more expensive, will have slightly less penetration than CO_2, and the finished weld bead has a smoother appearance.

FCAW-G with a mild-steel wire

Either CO_2 or C25 is used, depending on the chemical composition of the flux. When using C25 instead of CO_2, lower the voltage by 1 to 1.5 volts.

GMAW spray transfer with steel wires

All of the following gases are used with spray transfer. They are ranked by the amount of heat generated by the gas in the welding arc.

By law, cylinders need to have a label identifying the contents. This cylinder contains C25 used for short-circuit transfer with mild-steel wires. Do not use a cylinder if the label is missing or unreadable. *Monte Swann*

80 percent argon, 20 percent CO_2—very cool welding arc
90 percent argon, 10 percent CO_2—most commonly used
92 percent argon, 8 percent CO_2
98 percent argon, 2 percent CO_2
98 percent argon, 2 percent oxygen
95 percent argon, 5 percent oxygen—very hot welding arc

With stainless-steel wires

A mix of argon, helium (He), and CO_2 is used in various quantities, sometimes called a tri-mix shielding gas. A typical tri-mix gas would have 90 percent helium, 7.5 percent argon, and 2.5 percent CO_2. For spray transfer, 98 percent argon, 2 percent oxygen is commonly used.

With aluminum wires

100 percent argon used on materials up to ½-inch thick. An argon-helium mix is used on aluminum over ½-inch thick.

Crack open slightly, then close the cylinder valve to blow out any dirt or dust before installing a regulator. Support the regulator with one hand while threading it onto a cylinder. Use a tight-fitting wrench to make a snug connection. *Monte Swann*

Stand to the side of the regulator when opening a cylinder valve. The rush of high-pressure gas (up to 2500 psi) can cause pieces of the regulator to fly apart, especially if the regulator is old, worn, or damaged. *Monte Swann*

On a gauge-type regulator, set the flow rate in CFH by turning the adjustment screw clockwise while the gas is flowing. *Monte Swann*

On a glass-tube type regulator, use the adjustment knob to set the gas flow. The top of the ball indicates the amount of flow in CFH. Some regulator/flow meters have multiple scales for different shielding-gas mixes. Read the correct set of numbers for your shielding gas. *Monte Swann*

SHIELDING GAS FLOW RATE

Setting the proper flow rate for the shielding gas and wirefeed transfer mode is very important. A low shielding gas flow rate will not shield the molten puddle adequately and is more sensitive to any type of air or gun movement. A shielding gas flow rate set too high will cause air to be pulled in from the atmosphere to the weld pool by turbulence created in the high-pressure gas stream. The stability of the molten weld pool will also suffer from added turbulence.

For short-circuit transfer, a flow rate of 25–30 CFH is adequate.

For globular and spray transfers with solid and tubular wires, a flow rate of 35–45 CFH will be necessary. With these welding processes, the arc is larger, hotter, and more stickout is required when welding.

SETTING VOLTAGE AND WFS

There is an important relationship between voltage and wirefeed speed (amperage) in wirefeed welding. The two work in combination with each other, so if an adjustment is made with one, the other will probably need to be changed, as well.

There are two different ways of setting a wirefeed machine. One way is to set the WFS first (amperage) and then adjust the voltage second. I prefer setting the voltage first, then make the necessary adjustments to WFS to get the machine dialed in. Either method will work fine as long as the machine is set up correctly and the shielding gas matches the wire and the transfer mode.

Voltage influences the shape of the weld bead. Higher voltages will flatten out a weld bead. Lower voltages will make the weld bead more convex. If the toes of the weld are not adequately fused into the base metal, turn up the voltage. If the weld pool is too fluid, or the arc is burning through the base metal, turn down the voltage. In either case, an adjustment in WFS will also need to be made to maintain the proper ratio between volts and amps.

The voltage and wirefeed speed controls on this Lincoln SP-125 Plus are marked with letters and numbers. This makes it a little more difficult to set the machine exactly, without some trial and error. *Monte Swann*

This Miller CV has a digital display for open-circuit voltage and WFS in IPM. When welding begins, the display will change to indicate the working voltage. *Monte Swann*

An older machine, like this Hobart, may have meters to indicate amperage and voltage. The meter on the left for amperage reads zero and the OCV is shown. When welding begins, these meters will display working amperage and voltage. Notice the voltage adjustment knob has two sets of numbers for the high and low ranges. *Monte Swann*

The feeder control for the Hobart is a separate unit. The numbers around the adjustment knob are a percentage of maximum motor speed. For example, 10 = 100 percent and 5 = 50 percent. Use the technique for finding IPM to know the exact WFS for each setting. *Monte Swann*

GAS METAL ARC WELDING (GMAW)

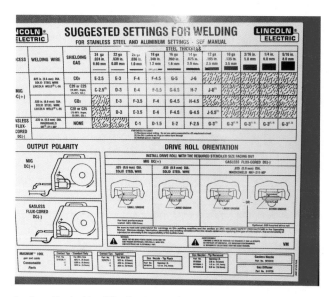

Look on the inside of the front cover for suggested machine settings, information on polarity, and drive-roll set up. *Monte Swann*

WFS influences the size of the weld bead. A high WFS will deposit a larger amount of filler metal in the joint. Low WFS will make a smaller weld bead. If an adjustment is made to WFS, voltage will also need to be adjusted in order for the machine to run properly. Remember that as you increase the WFS, you also increase the amperage and the amount of heat.

Between 15 and 32 load volts, the ideal ratio between amps and volts is **6:1**—for every one volt of load, six amps are needed to burn off the wire properly. For example, if you set the machine for 20 volts, multiply that number by 6 for the correct amperage: 20 × 6 = 120 amps. There is a fair amount of leeway in setting the machine, but this 6:1 ratio will get you close to the correct settings from the start.

For mild-steel sheet metal, a setting of one amp for every 0.001 inch of material thickness can be used. For example, if you are welding 14-gauge mild steel, the thickness is 0.074 inch, so, around 74 amps can be used as a starting point. Some welders will set the required voltage for the mode of transfer and start with 50 percent of the maximum WFS, making test welds to dial in the machine.

If your machine has a display for working amps and volts, setting the machine is easy. If the machine gives WFS in inches per minute, or as a percentage of wirefeed motor speed, chart 11-2 shows a method that can be used to determine amperage for mild-steel wires.

Let's go through an example of setting up a wirefeed machine. You are going to weld ⅛-inch thick mild steel and you need a place to start.

1. Convert the material thickness to amps, ⅛ inch = 0.125, so 125 amps would be a good place to start.
2. Our basic ratio between amps to volts is 6:1. 125 amps/ 6 volts gives us a machine setting of around 21 volts.

3. From Chart 11-2, for 0.035-inch-diameter wire we should start with an IPM-to-amps ratio of 1.6:1. To calculate WFS, multiply 1.6 times the number of amps. 1.6 × 125 = 200 IPM WFS.
4. We can estimate the IPM setting by pulling the trigger for 6 seconds, measuring the wire, and multiplying by 10.

Experienced welders may balk at all this math used to set amps and volts. Through years of experience they have learned to dial in the machine by the sound of the arc and the appearance of the weld bead. Some welders can tell if a machine is set right in the few seconds it takes to make a tack weld. Listen to your machine and experiment with different machine settings. Once you find a good setting where the machine is running smoothly, write it down. Next time you need make a similar weld, you will not need to reinvent the wheel. You can even take this idea a few steps further and keep a running journal of how each machine you use is set up for every type and thickness of metal you ever successfully weld.

There can be many reasons why a wirefeed machine is not running right or the weld bead is poor. Not having the correct voltage and WFS is just one of the possible causes. See the section on troubleshooting wirefeed problems for more information. Make sure the machine is set up correctly before trying to make fine-tune adjustments with the controls.

SETTING WIRE TENSION

The welding wire needs to feed properly through the system to prevent problems. Drive-roll tension should be set for the pressure required to feed the wire correctly through the gun cable and to the work. Too much drive-roll tension will pinch the wire between the rollers and could cause birdnesting. If knurled rollers are used, excessive pressure on the wire will put tooth-like indentations in the wire. As the deformed wire

GMAW conversion chart for mild-steel electrodes

To determine WFS in IPM:
1. Cut off the end of the wire
2. Pull the gun trigger for 6 seconds
3. Measure the length of wire fed in that time
4. Multiply the length by 10 for the IPM

Use the following chart to determine amperage

Wire Diameter	IPM	AMPS
0.045 inch	1	1
0.035 inch	1.6	1
0.030 inch	2	1
0.023 inch	2.4	1

Chart 11-2

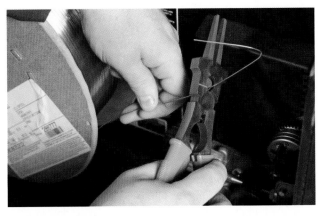

Keep the wire threaded through the side of the spool until you need it. Cut off the bent end before feeding it through the drive rolls. Keep a good grip on the wire; if you let go, it will quickly unspool on its own. *Monte Swann*

With the tension screw released, thread the wire through the inlet guides and center it over the groove before securing the top roller. Notice these drive rolls have two grooves. Some drive rolls are reversible and will work with two different wire diameters. *Monte Swann*

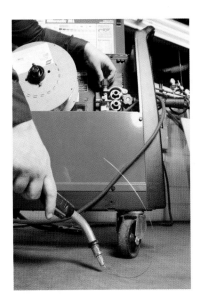

To adjust drive-roll tension, feed the wire against a hard surface. If the wire stops feeding, tighten the tension screw until wire slippage is eliminated. *Monte Swann*

is fed through the liner and contact tube, the indentations act like a saw, cutting into the consumables, quickly wearing them out.

If tension is set too low, wire slippage will cause irregular feeding, burnback of the wire onto the contact tip, or the wire could stop feeding altogether. Set drive-roll tension just until wire slippage is eliminated.

Spool-brake tension is another important consideration. It should be set so that the wire will not continue to unspool after the drive rolls have stopped turning. Too much tension will make the drive rolls pull harder than they have to, which could damage the motor. To check spool-brake tension, release the tension adjustment screw on the drive rolls. Use a pair of pliers to pull the wire manually through the end of the gun. The wire should be easy to pull without too much effort. If the wire is difficult to pull, check the spool-brake tension. If it is not set too tight, then the liner may be clogged or kinked and will need replacing.

TROUBLESHOOTING

Since there are many parts of a wirefeed system, more things can go wrong. Each wirefeed problem can have multiple causes. Troubleshooting wirefeed problems is a useful skill. Effective troubleshooting requires some knowledge of the welding process and machine being used. It can seem complicated and overwhelming at first. But with time and experience, troubleshooting wirefeed problems will not be difficult.

Adjust your spool-brake tension for proper feeding. When feeding stops, there should be some slack in the wire without it over spooling. Do not overtighten the spool brake. *Monte Swann*

Clearing a birdnest like this one is easy. Cut the wire off and secure it on the spool. Remove the contact tip from the gun. Release the drive-roll tension screw and pull the wire out of the gun cable liner. Coil it up since long pieces of wire left on the floor are a tripping hazard. If birdnesting is a constant problem, take a few minutes to figure out the cause. *Monte Swann*

Birdnesting can be caused by:

Drive-roll tension-adjustment screw too tight
Drive rolls are the incorrect size for the wire diameter being used
Drive rolls are misaligned with each other, worn down, or damaged
Contact tip or gun cable liner is the incorrect size for the wire diameter
Gun cable is kinked or coiled too tightly during welding
Gun cable is too long for the drive-roll system
Restrictions in the gun or gun cable
Wirefeed set up is not correct for softer wire, such as stainless steel or aluminum

Irregular wirefeed can be caused by:

Drive-roll tension too lose
Gun cable liner kinked, clogged, or worn out
Kinked welding wire (electrode)

If you think there is a kink or obstruction in the liner, check it by feeding the wire through the gun, first. This only works if the machine has an inch control. *Monte Swann*

Erratic arc can be caused by:

Dirty weld joint area
Loose contact tip
Contact tip plugged or worn
Shielding gas flow set too low
Poor work-clamp connection to paint, rust, or dirt
Incorrect polarity
Lose cable (welding lead) connections
Incorrect voltage and WFS settings
Separation of mixed gases. CO_2 is in a liquid form in the cylinder. Liquid CO_2 can settle over time. To remix the gases, put the cylinder cap on and roll the cylinder on the ground. Keep in mind the ratio of mixed gases is only good down to 200 psi. Below that cylinder pressure, the gas mix may be different.
Incorrect shielding gas being used for the type of wire or mode of transfer
Hole in gas hose or lose connections aspirating air into the shielding gas lines

Porosity in a wirefeed weld is always the result of contamination. Atmospheric contamination from a lack of shielding gas will make the weld bead brittle since oxygen, nitrogen, and hydrogen are absorbed by the molten weld pool. Surface or subsurface porosity and crater cracks can be caused by:

Dirty base metal or contaminated electrode
Drafts or windy conditions blowing shielding gas away
Gas flow set too low
Gas flow set too high
Moisture in the shielding gas
Incorrect shielding gas being used for the type of wire or mode of transfer
Gun too far away from weld pool
Contact tube recessed too far into nozzle (contact tube setback)
Contact tube not centered in nozzle
Travel speed too high
Empty gas cylinder
Kinked gas hose or obstruction in gas lines
Leaks in the gas lines, including the regulator and gun
Broken gas solenoid

WELDING TECHNIQUE

Although less skill is required for wirefeed, the weld is still affected by the four main variables: distance, angle, speed, and heat.

Distance

Wirefeed welding machines are self-regulating. This means the distance between the gun and base metal does not have to be exact in order to make a good weld. This distance is called stickout, which is the distance from the contact tube to the work (electrode extension/actual stickout) or from the nozzle to the work depending on how it is measured. As stickout increases, preheating in the welding wire also increases, so the

At certain angles, the nozzle will block your view of the molten weld pool. This makes it difficult to keep track of the joint. Try to keep your head positioned so you can see the end of the wire fusing into the base metal. Move your head—not the gun—to maintain the proper viewing angles. *Monte Swann*

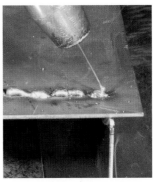

Too much stickout will compromise shielding-gas flow and create an uneven bead. Notice the inductance of the welding current is heating up the wire until it glows orange and balls up on the end. Welding this way reduces the amount of heat in the base metal. *Monte Swann*

With the WFS set too low, the travel speed is slow and the wire is likely to burn back onto the contact tip. *Monte Swann*

When WFS is set too high, there is not enough voltage to fuse the large amount of filler metal into the base metal. The wire is stubbing out and it feels like the gun is being pushed back in my hands. *Monte Swann*

machine will automatically adjust by lowering the amount of current. As stickout decreases, the machine will increase the amount of amperage to compensate for the increased rate of wire burn-off. Either way, the machine is making these slight adjustments at a very fast rate.

There are limits to stickout. Too much stickout increases voltage and lowers the amperage, which tends to destabilize the arc, causing excessive spatter, poor heat control, and insufficient penetration. A long stickout also puts an excessive amount of preheat into the wire, increasing the rate of filler metal deposited into the weld, resulting in a lack of fusion between the base and filler metal. Too little stickout will result in the wire burning back and fusing to the end of the contact tip.

Gas Nozzle

Contact Tube

Contact Tube Setback

Electrode Extension

Workpiece

Standoff Distance

Stickout

Consumable Electrode

Welding Inspection Technology, Fourth Edition, Figure 3.15, Reproduced with permission from the American Welding Society (AWS), Miami, FL USA

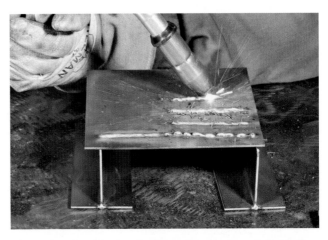

When voltage is set too low, the filler metal cannot penetrate into the base metal. The bead has a convex contour and lack of fusion at the toes. *Monte Swann*

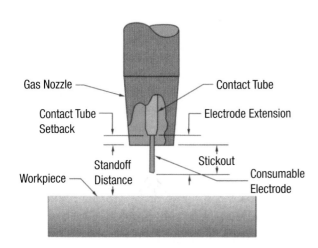

The amount of stickout can vary depending on the wirefeed process:

- Short-circuit transfer; stickout of ¼ to ½ inch is used—⅜-inch stickout is ideal.
- Spray, pulse spray, and FCAW-G; stickout of ⅝ inch minimum to ¾ inch maximum.
- FCAW-S; stickout of ½ inch minimum to 1 inch maximum.

Contact tube setback

Another important distance consideration is the contact tube setback or contact tip recess. Contact tip recess affects the amount of stickout: the greater the recess, the longer the wire stickout. Typically, the contact tip should be kept flush or slightly recessed into the nozzle, about ⅛ inch. A ¼-inch recess is used in spray and pulse-spray transfers above 200 amps. A ⅛-inch contact tip extension can be used with short-circuit transfer (below 200 amps) for welding difficult-to-access joints. Keep in mind flush and extended contact tips position the nozzle farther away from the molten

I have purposely turned the shielding gas off. The arc is much brighter because the oxygen in the air is increasing the amount of heat in the arc and contaminating the weld bead. Notice the large amount of fumes and smoke generated when the arc is exposed to the atmosphere. *Monte Swann*

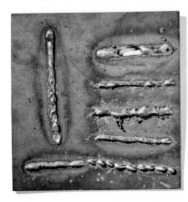

The example welds from top to bottom:
- Voltage set too high (burn through)
- Voltage set too low
- WFS set too high
- WFS set too low
- Too much stickout (too far away)

The bead on the left side was made without shielding gas. *Monte Swann*

puddle. This can compromise the shielding gas flow, making the weld susceptible to any air movements or contamination from the atmosphere.

Relationship to *angle*:

Steep gun angles increase wire stickout, especially if access to the joint is limited or the nozzle interferes with the distance from gun to the base metal.

Relationship to speed:

Faster travel speeds can be used with a shorter stickout.

Slower travel speeds are required with a longer stickout.

Angle

Gun angle directs the heat of the welding arc, deposited filler metal, and shielding gas flow. Radical gun angles cause undercut, and compromise the flow of shielding gas. Proper work and travel angles are different depending on the joint design and welding position. Move your head to different locations until you have a clear view of the end of the wire feeding into the molten puddle.

Speed

Two different speeds are involved in wirefeed. The WFS is set on the machine in IPM. Travel speed is determined by the person manipulating the gun. Overall, travel speeds are faster with wirefeed welding. It is easy to miss the joint with wirefeed, because everything happens quickly. Use the end of the wire as a guide and keep an eye on where the weld pool is going. A fast travel speed will result in a small weld bead and lack of fusion into the base metal. A slow travel speed will build up too much heat and filler metal in one area. Small circular or zigzag motions can be used for wirefeed welding. Keep all motions small and tight; sudden movements can cause undercut, or incomplete fusion.

Heat

The heat required for welding is determined by the amount of amperage at the welding arc. Since a CV power source is used in wirefeed welding, the level of voltage can be adjusted. It may help to think of adjusting the heat level by making changes to the voltage setting first, and then adjusting the WFS to match the voltage being used.

As the amount of voltage increases, different modes of metal transfer take place. These different modes put different amounts of heat to the base metal. Too much voltage can cause burn through. High levels of heat also cause unnecessary distortion in the work piece. Too little voltage will result in a lack of fusion between the base metal and filler metal. Lack of fusion causes a welded joint to come apart under load.

Relationship to *speed*:

Within the individual modes of transfer:

Faster travel speeds are used with more heat

Slower travel speeds are used with less heat

Another way to measure heat input is by determining the joules of energy required for a weld. In industry, some welding procedures require a certain number of joules. Changing WFS, volts, and travel speed will change the heat input. For your reference, use the following formula to determine heat input in welding.

$$\text{Joules of energy} = \frac{\text{Amps} \times \text{Volts} \times 60}{\text{IPM (or travel speed)}}$$

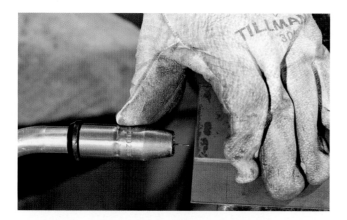

To steady your gun while tacking, use your thumb; place the nozzle on the work or rest your arm on your leg. *Monte Swann*

Push vs. pull

A wirefeed gun can either be pushed—pointing in the direction of travel—or pulled (dragged) in the opposite direction of travel. Pulling a wirefeed gun (backhand technique) results in deeper penetration into the base metal and a convex weld bead. Pulling is primarily used for welding thicker metals with short-circuit transfer. When welding is done in tight spaces, there sometimes are joints that can only be welded by pulling the gun.

Pushing the gun (forehand technique) will flatten out the weld bead and results in less penetration. Pushing is used to weld thinner pieces of metal with short-circuit transfer. A pushing technique is also used when welding with spray or pulse-spray transfer.

Starts and restarts

Starting a bead with wirefeed is easy. Get the gun positioned first, place the wire over where you want to begin, then pull the trigger. Cutting the wire off at an angle will help to start the arc quickly. When restarting a bead, direct the wire ½ inch ahead of the end of the last bead (crater). Come back over the top of the crater, circle the gun once, and continue welding. With a FCAW process, chip off the slag at the end of the previous bead before restarting to prevent incomplete fusion or slag inclusions.

When ending a bead, let off the trigger before pulling the gun away. The gun should be kept over the end of the weld for one second after welding to let any additional shielding gas protect the solidifying weld bead. If the gun is yanked away abruptly from the weld pool, a porous crater will be left, and possibly an arc strike.

Tacking technique

Tacking pieces together is easy with wirefeed. The biggest problem beginners have is they miss with the wire, usually leaving a mess of wire or weld bead outside of the joint area. Support the gun or nozzle with your hand or body to steady the gun before pulling the trigger. The nozzle, and even the end of the wire, can come in contact with the base metal before welding begins. The nozzle is insulated against the welding current and the wire is not electrically hot until the trigger is pulled on the gun.

It's easy to make cold tacks with a wirefeed machine. If the tack is not fused into the base metal, it will not hold the joint in alignment during welding. Double-check the tacks to verify that there is adequate fusion into the base metal. Some welders will use their eyelids as a shade, closing their eyes or looking away while making tack welds. Remember, the light from the arc will still burn any exposed skin.

SETUP

Become familiar with all the controls on the machine you are using. Although wirefeed welding is easier than other processes, it can be very frustrating when the machine is not working right. Take a few minutes to experiment with different voltage and amperage settings. See what happens when each one is set too high or too low. Try turning off the shielding gas. Know what it looks and sounds like when your molten weld pool is in direct contact with the atmosphere.

Set up the machine for your project by:

1. Selecting the type of wire based upon the composition of the base metal and service conditions for your project. Machines or parts used in extreme heat, cold, or exposed to corrosive elements may require a specialized filler metal.
2. Selecting the wire diameter based on your machine's output capacity and base metal thickness.
3. Matching the shielding gas to the type of wire and mode of transfer.

Note: wirefeed welding machines may only work with certain sizes of wire and specific modes of transfer.

Get set up right for the job you are doing. No matter how many years of experience a welder may have, there is no possible way for him to make a good weld on aluminum using CO_2 as the shielding gas. A less experienced person may blindly change machine controls in the hope of stumbling upon a setting that works. Not knowing welding and machine basics is like asking an artist to paint a portrait with a blindfold on.

Steady your body by resting your arms on the table or bracing yourself against a solid object. Use two hands on the gun. Relax and get comfortable so the gun can be guided in a smooth and uniform motion. Pay attention to the end of the wire feeding into the weld pool. If you are having a hard time seeing, adjust the position of your head—not the gun angle—so you will have a clear view. With short-circuit wire transfer, the wire is small and the arc is less bright. Use cheater lenses or bifocals and switch to a No. 9 shade if needed.

FCAW

Although this chapter focuses on GMAW, most of the information on machine controls, equipment, troubleshooting, and welding technique is the same for flux core arc welding (FCAW). The only additional tools needed for FCAW are a chipping hammer and wire brush to clean off the slag.

The flux in FCAW serves the same purpose as in SMAW:

- Acts as a cleansing and deoxidizing agent in the molten weld pool.
- Creates an inert atmosphere by releasing CO_2, shielding the molten pool from contamination. With FCAW-G, a shielding gas is necessary to fully protect the weld.
- Forms a slag coating over the semi-molten metal to further protect it from the atmosphere.

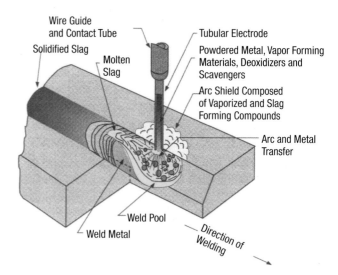

Welding Inspection Technology, Fourth Edition, Figure 3.16, Reproduced with permission from the American Welding Society (AWS), Miami, FL USA

This is a weld using a self-shielded, flux-cored electrode. The wire diameter is 0.035 inch, which is a small size for this welding process, operating in a short-circuit voltage range. Larger amounts of fumes, smoke, and spatter are generated by FCAW-S. *Monte Swann*

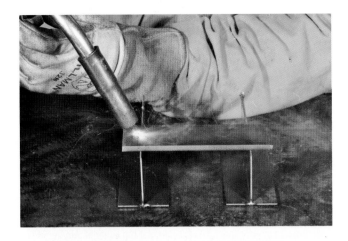

Using between 22 and 24 load volts, this 0.045-inch-diameter flux-core wire penetrates deep into the ⅜-inch-thick base metal. The flux inside the wire is supplemented with a shielding gas in the FCAW-G process. *Monte Swann*

- Stabilizes the welding arc.
- Improves penetration (depth of fusion) into the base metal.
- Determines type of welding current used: DCEP or DCEN.

All of the joint designs and welding positions can be used as exercises for FCAW.

WELDING EXERCISES

I recommend completing these exercises in the order presented. The level of difficulty increases with each exercise, so become proficient with your new skills in one exercise before moving to the next. Wear the proper PPE (see chapter 2 for more information).

For the GMAW exercises you will need:

A CV wirefeed welding machine

ER 70S-6 wire (0.035 diameter was used)

C25 shielding gas (CO_2 can also be used)

14-gauge, hot-rolled, mild-steel pieces cut 2 × 6 inches

The slag byproduct from the flux is removed with a chipping hammer and wire brush. On top is the FCAW-S and below is the FCAW-G bead. Both still have some flux residue on the bead. I may be able to reduce the amount of spatter on the FCAW-G bead by increasing the amount of stickout. *Monte Swann*

These two pieces of ⅜-inch metal are joined using FCAW-G. The first pass on the open-root corner joint is penetrating all the way through. At the beginning, the arc cut into the top edge, causing some of the molten filler metal to drip out. Notice a longer stickout is used. *Monte Swann*

After cleaning the flux off the root pass, I made a second pass to fill the joint. It will be easy to remove the extra weld metal that flowed out of the joint on the first pass. *Monte Swann*

EXERCISE 1
Bead on a plate
Gun angles:

      Work angle: 90 degrees

      Travel angle: 55 degrees

There are two reasons for doing this exercise: first to get the machine setting dialed in correctly, and second to run practice beads on the same thickness of metal as your project. Make adjustments to the voltage and WFS until the amount of heat and bead profile looks correct. Any problems with gas flow and the wirefeed system can be fixed, as well. Once the machine settings are dialed in, focus on welding technique. Experiment with different gun angles and travel speeds.

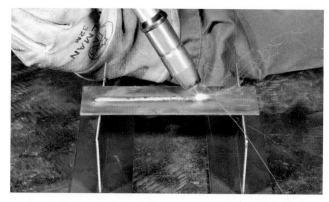

Keep the wire on the leading edge of the puddle. Travel in a straight line, or make loosely spaced, small, circular movements. *Monte Swann*

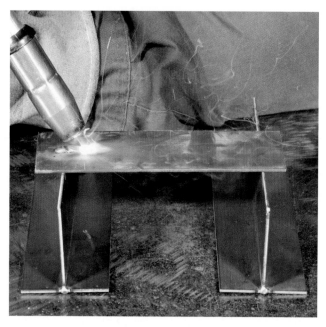

Once you pull the trigger, start moving. As long as the wire is feeding, weld bead is building up fast. *Monte Swann*

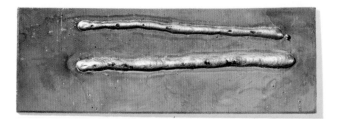

Each of these beads was made with the same machine settings, but at different travel speeds. *Monte Swann*

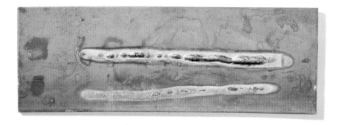

The larger bead, made at a slower travel speed, melted through the back side of the plate. In order to travel slowly one needs to turn down the voltage and WFS to prevent excessive penetration or burn-through on thinner sheet metals. *Monte Swann*

EXERCISE 2
Lap joint, flat position
Gun angles:

      Work angle: 55 degrees from the base plate (upper plate)

      Travel angle: 45 degrees

A lap joint can be tacked at each end so no tacks are left in the joint. Overlap the pieces halfway, leaving enough of a space on the base plate. A narrow space will concentrate heat quickly and may lead to melt-through. The edge of the lap joint should be consumed by the weld bead. There should be a smooth transition between the toe of the weld and the base plate, without overlap or lack of fusion.

Maintain a ⅜-inch stickout during the entire weld. *Monte Swann*

Rest one arm on the table and keep both hands on the gun in order to remain steady and have good control while welding. *Monte Swann*

The finished fillet weld should have good fusion to the base plate. The weld should only come up to the edge and not overlap the top plate. *Monte Swann*

Keep the wire directed at the base plate, a little above the root of the joint. Use a small circular motion to wash the molten puddle up on the edge. *Monte Swann*

EXERCISE 3
Tee joint, flat position
Gun angles:

Work angle: 45 degrees from each plate

Travel angle: 45–55 degrees

Start by tacking each end of the tee joint so the two pieces are at a 90-degree angle. Use a scrap piece as your base and tack the tee joint at a 45-degree angle; this way, the fillet weld you are making is in the flat position with the weld facing straight up. Try this exercise without a root opening in the joint. For an additional challenge, add a 1/16-inch gap between the pieces. A tee joint with a root opening will have a groove weld applied to it. Either way, a tee joint requires more voltage than a bead on a plate or lap joint because the volume of the surrounding metal rapidly absorbs heat away from the weld area.

Push the weld along the joint. Pay attention to where the wire is feeding. It is easy to get off track and weld up onto the corner. *Monte Swann*

It may be difficult to maintain proper stickout with a tee joint. The nozzle of the gun may be too large to get close enough in tight places. Keep the contact tip flush with the nozzle to minimize the amount of electrode extension. A smaller-diameter nozzle can also be used. *Monte Swann*

Use small, circular motions to wash the weld pool up onto both plates. Notice how close the nozzle is to the sides of the base metal. *Monte Swann*

The finished fillet weld has adequate fusion at the toes and equal legs. Notice part of the bead is washed up on the top plate. Any small movement of the gun is visible in the final weld. *Monte Swann*

As you circle around, bring the wire back to the leading edge of the pool for fusion at the root of the joint. *Monte Swann*

I may have pulled the gun away too soon. The end of the bead is still very hot and could have benefited from some more time protected by the shielding gas. *Monte Swann*

EXERCISE 4
Square groove butt joint with ¹⁄₁₆-inch root opening, flat position

Gun angles:

Work angle: 90 degrees

Travel angle: 45–55 degrees

Direction of travel has a great effect on the bead profile and amount of penetration. I began this exercise by pulling the gun, without good results. Next, I pushed the gun and was able to carry the bead across the joint without burning through. Experiment with this exercise by using different root openings and metal thicknesses. This exercise may require less voltage than a tee or lap joint.

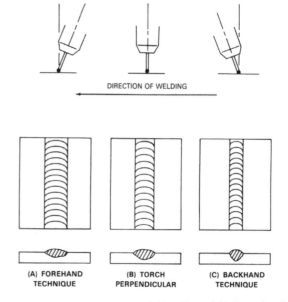

DIRECTION OF WELDING

(A) FOREHAND TECHNIQUE

(B) TORCH PERPENDICULAR

(C) BACKHAND TECHNIQUE

Welding Handbook Volume 2, Eighth Edition, Figure 4.13, Reproduced with permission from the American Welding Society (AWS), Miami, FL USA

For an experiment, I'm going to pull the gun using a backhand technique. *Monte Swann*

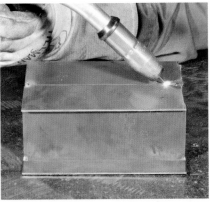

Everything is working well at the beginning of the weld. *Monte Swann*

A couple of inches in, the wire is burning through the sheet metal. Even tilting the gun down to decrease the travel angle is not enough to compensate for the increased penetration. *Monte Swann*

On another joint I'll try the forehand technique, pushing the gun. *Monte Swann*

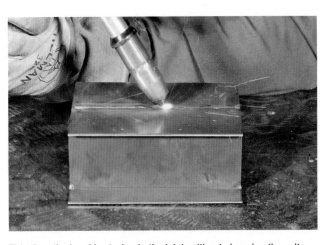

This time the bead is staying in the joint, without changing the voltage or WFS. If the weld did burn through, decrease the volts and WFS. *Monte Swann*

Keep the wire directed on the leading edge of the molten pool and centered over the root opening. There is no need to use a circular motion. Let the wire fill in the joint by using a smooth, steady travel speed. *Monte Swann*

EXERCISE 5

Tee joint horizontal position

Gun angles:

Work angle: 35–40 degrees

Travel angle: 45–55 degrees

This exercise works about the same as a tee in the flat position. The gun angle is tilted down slightly to compensate for the effect of gravity on the molten weld metal.

Use two tee joints to prop up the base metal. Start the weld with a clear view of the entire joint. *Monte Swann*

Use a work angle of 35 degrees to compensate for the effects of gravity. Direct the welding wire so enough filler metal washes up onto the top plate. *Monte Swann*

The finished fillet weld should be uniform and consistent with equal legs. *Monte Swann*

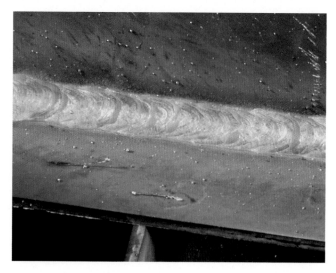

You may notice some brown spots on the finished weld bead. This is silica and other impurities, which have floated to the top of the weld pool. These are easily removed by chipping or using a wire brush. *Monte Swann*

EXERCISE 6

Square groove butt joint with $\frac{1}{16}$-inch root opening, horizontal position

Gun angles:

Work angle: 80–85 degrees

Travel angle: 45 degrees

Tack up the joint like you would for welding it in the flat position. Tack the joint to another piece of metal or use a fixture to support the pieces in the horizontal position. Although the metal is vertically aligned, the seam of the joint is positioned horizontally. Keep the gun tilted down slightly (nozzle pointed up) to compensate for the effects of gravity. By positioning the joint at the edge of a work table, you should be able to hold the gun at a comfortable angle.

Begin the groove weld with the wire centered over the root opening. The gap between the plates should be evenly spaced. *Monte Swann*

It is not necessary to use a circular motion for this weld. Use a work angle slightly less than 90 degrees. Push the bead along the joint; the bead may sag a little from the effects of gravity. *Monte Swann*

Some parts of the bead are wider than others—a result of inconsistent travel speed. Maintaining a consistent travel speed is more difficult when the weld joint is up off the table. *Monte Swann*

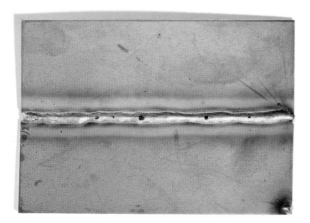

The finished weld is not completely uniform, but both pieces are welded together with good fusion at the toes and adequate penetration into the joint. *Monte Swann*

EXERCISE 7
Square groove butt joint with ¹⁄₁₆-inch root opening, vertical position, downhill

Gun angles:

    Work angle: 90 degrees

    Travel angle: 45 degrees – pull

A great way to make a smooth, flat weld bead on a joint is to travel vertical down. Typically, with short circuit transfer, thinner gauge metals are welded. Traveling downhill reduces the amount of penetration, making it easier to weld thin metals. CJP can be achieved on 14-gauge sheet metal when welding downhill. Metals over ⅛-inch thick can be welded traveling uphill for better penetration.

Notice a pulling technique is used for travel direction. There is less penetration when the molten weld pool is moving with gravity while traveling downhill. Because we are using short-circuit transfer, the weld pool solidifies quickly. The relationship between heat and root opening is important. As a general rule, the root opening should not exceed the base-metal thickness. An overall lower voltage should be used for welding vertical down with an evenly spaced gap between the pieces.

      (A) Downhill   (B) Uphill

Welding Handbook Volume 2, Eighth Edition, Figure 4.15, Reproduced with permission from the American Welding Society (AWS), Miami, FL USA

I've positioned this butt joint near the edge of the table, allowing me to hold the gun at a comfortable angle. The 45-degree neck of the gun (called a goose neck) is correct for the travel angle. *Monte Swann*

Pull the gun straight down the joint. As you weld, don't let the molten pool above the wire become too large. With the right travel speed, the liquid pool will follow the welding wire and not sag down. It is easy to burn a hole through the joint if traveling too fast. *Monte Swann*

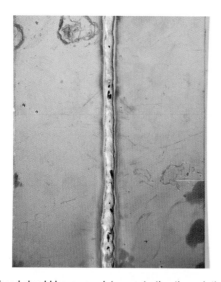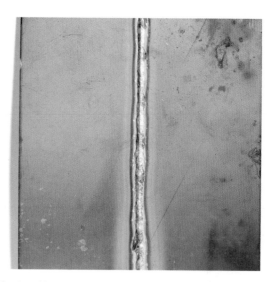

The finished bead should have complete penetration through the joint. Both the front and back side look like they have been welded, even though the joint was welded from one side only. The fast-freeze action on short-circuit transfer makes this type of wirefeed weld possible. *Monte Swann*

For an additional challenge, bend angles in the base metal, either 45 or 90 degrees; practice welding around outside and inside corners. It takes some skill to be able to turn the corners and make these welds without stopping. *Monte Swann*

CHAPTER 12
GAS TUNGSTEN ARC WELDING (GTAW)

GTAW

Gas tungsten arc welding (GTAW) is often referred to as tungsten inert gas (TIG) welding. Unlike wirefeed, there are not as many variations in the names used to describe GTAW. The process is similar to gas welding in that a molten puddle is created on the base metal before a filler metal is added. Unlike the oxy/acetylene flame, the GTAW arc is much hotter. With the intense, focused arc, heat can be directed with enough accuracy to weld hypodermic needles together. Let's take a look at the acronym "GTAW":

Gas

A shielding gas is used to protect the molten weld pool from atmospheric contamination. The air we breathe contains elements such as nitrogen, oxygen, and hydrogen, which molten metal can rapidly absorb. If these elements are absorbed, the weld will be brittle and porous. In GTAW, a shielding gas is always used. A flux is never part of the electrode, but flux-coated or flux-cored filler rods are sometimes used.

Tungsten

The type of electrode sets GTAW apart from the other arc welding processes. The electrode is made of tungsten.

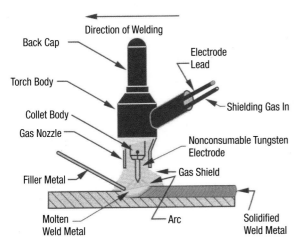

Tungsten is used to carry the arc, which creates a molten puddle. A filler rod must be added to build weld metal in the joint. Since the tungsten does not become part of the finished weld bead, it is a non-consumable electrode.

Arc

Electricity flows through the welding circuit. Resistance is created when the electrons cross the gap between the tungsten and base metal. The resistance in the welding circuit creates the heat required to melt the base metal.

Welding

Either autogenous welds (made by fusion of the base metals without filler) or welds reinforced with a filler metal can be made with GTAW. Of all the manual welding processes, GTAW can join the widest variety of metals with the least amount of changes to the equipment.

GTAW MACHINES

Any CC power source can be used for GTAW. Some CC machines are specifically designed for GTAW with either AC or DC, providing the most versatility in welding. This chapter will focus on welding with a machine designed for GTAW. See chapter 9 for more information on welding machines.

MACHINE CONTROLS
Polarity AC/DC

Most TIG welding is done with DCEN power. With DCEN, the electricity arcs from the tungsten electrode to the base metal. This concentrates a majority of the heat on the base metal, creating a molten weld pool. Steel, stainless steel, copper alloys (like brass), nickel alloys, silver, gold, and titanium are all welded with DCEN. AC is used to weld aluminum and magnesium. If a power source is not designed for AC TIG welding, it will be difficult to weld aluminum and magnesium.

DCEP power is rarely used in TIG welding. With DCEP, the electricity arcs from the base metal to the tungsten. This concentrates a majority of the heat on the tungsten electrode, causing it to melt away. When welding with DCEP, use a larger diameter electrode tapered to a point, and a lower amperage setting.

Welding Inspection Technology, Fourth Edition, Figure 3.20, Reproduced with permission from the American Welding Society (AWS), Miami, FL USA

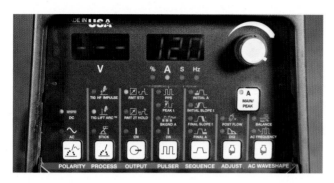

This inverter power source can be used with either AC or DC to TIG weld almost any type of metal. It also has features such as high-frequency start, pulser, sequencer, and controls to change the frequency and balance of the AC wave. The machine is set for 120 amps; this is the maximum output when using a remote amperage control. *Monte Swann*

Symbols of a work clamp and torch are used on this power source to indicate the positive and negative terminals. The work clamp is connected to the positive, and torch connected to the negative for DCSP. The center outlet is marked with the number 14. This is for a 14-pronged plug used to hook up the remote amperage control to the machine. *Monte Swann*

Amperage

GTAW is similar to SMAW. Amperage is set on the machine, and voltage is determined by the arc length. Amperage is used to adjust the amount of heat required to melt the base metal.

High frequency

There are several ways to start a TIG welding arc. Some GTAW machines will have a high-frequency (HF) start control. A compressed AC wave is used to make the electricity arc between the tungsten and base metal. With HF, an arc can be started without the tungsten coming in contact with the base metal. Machines may have a switch giving the options for HF start, HF continuous, and HF off. HF start is used for TIG welding with DC. Since the welding arc is destabilized when AC is used, the HF needs to be in continuous mode. Newer

machines change the high-frequency setting automatically, depending on which type of current is selected. An SMAW/CC machine with AC polarity cannot be used for AC GTAW welding. In order to use AC polarity for TIG welding aluminum and magnesium, the machine has to supply continuous HF.

Remote amperage control

Most GTAW machines will have the option of controlling the amperage remotely. Either a foot pedal or torch-mounted switch is used; foot pedals are more commonly used. Torch-mounted switches are useful when welding in difficult-to-reach places, where using a foot pedal is impractical. The remote control increases or decreases the amount of amperage output coming from the machine. This gives the operator a great leeway over

Current Type	DC	DC	AC (Balanced)
Electrode Polarity	Negative	Positive	
Electron and Ion Flow			
Penetration Characteristics			
Oxide Cleaning Action	No	Yes	Yes - Once every half cycle
Heat Balance in the Arc (approx.)	70% At work and 30% At electrode end	30% At work and 70% At electrode end	50% At work and 50% At electrode end
Penetration	Deep; narrow	Shallow; wide	Medium
Electrode Capacity	Excellent (e.g. 3.18 mm [1/8in] - 400 A)	Poor (e.g. 6.35 mm [1/4in] - 120 A)	Good (e.g. 3.18 mm [1/8in] - 225 A)

Welding Inspection Technology, Fourth Edition, Figure 3.21, Reproduced with permission from the American Welding Society (AWS), Miami, FL USA

The foot pedal is a common remote control for amperage. Keep your floor swept and the area clean when using a foot pedal. Otherwise, dirt and metal dust can easily get inside the foot pedal and damage the components. *Monte Swann*

the heat input to the base metal. Remember: In arc welding, the amount of amperage controls both the rate of heating and the rate of cooling.

With HF, the arc is also started by remote control. Machines without remote amperage control can still be used, but the amperage is set only at the machine. Without a remote amperage control, once the arc is started, the maximum amperage is maintained during the weld. The amperage set on the machine has to be more exact, and a uniform travel speed is critical.

Lift start

With a lift-start control, the tungsten electrode is touched to the base metal and then lifted slightly to begin the arc. Remote amperage control can be used with the lift-start option. Only DC is used with a lift start.

Preflow/postflow

The shielding gas flow can be set before the arc is started (preflow) and/or after the arc has stopped (postflow). Preflow floods the weld area over the base metal with a protective atmosphere before the arc is even struck. It is not necessary to use preflow in most TIG welding applications. Postflow is a more important function, because it continues to shield the molten weld pool as it solidifies at the end of a weld. Postflow also protects the white hot tungsten electrode from contaminating elements in the atmosphere.

The amount of time needed for postflow will depend on the diameter of tungsten being used. Set 1 second of postflow time for every 0.010 inch of tungsten. For example, if you are using a 3/32-inch-diameter tungsten electrode, you first have to convert the fraction to a decimal. The fraction 3/32 = 0.093, so set the postflow between 9 and 10 seconds.

Pulse

Some GTAW machines use specific controls to fine tune the amount of amperage. Pulsing controls can be set so the amount of amperage varies between a high value (peak) and low value (background). As the current hits its peak value, a molten puddle forms on the base metal and filler metal can be added. As the amperage changes to its background value, the base metal cools and the molten pool solidifies or is in a state between liquid and solid. The amount of time and level of current can be adjusted according to specific needs. Using a pulse control greatly reduces the amount of heat in the base metal during welding. Pulsing can be accomplished by using a remote amperage control, but the timing and heat input will be much less accurate.

Slope time (sequencer)

Slope time controls the amperage at the beginning or end of the weld. At the beginning, the base metal is at the lowest temperature during the weld. Amperage can be increased temporarily, allowing the metal to heat up. At the end of a weld, the metal is at the highest temperature, so the amperage should be slowly reduced. This feature is called crater-fill on some machines. Crater-fill maintains lower amperage required to fill the crater at the end of a weld, without the molten pool becoming too fluid. Slope time is used to regulate the amperage when using a machine without a remote amperage control. Sequencers are used to program slope times and amperage settings for TIG welding with automated equipment.

AC frequency

AC is supplied in a wave form, alternating between the positive and negative electrical poles. As the current crosses from one pole to the next, the arc is destabilized. The arc can also be extinguished when the polarity reaches the zero value. As it crosses to each pole, current rises in a slope to a maximum value at the positive or negative pole. This creates an amperage lag. To get around this lag in current flow, square-wave technology is used to make the transition between the poles in a much shorter time. The square wave brings the current up to maximum value almost instantly. A lower-amperage HF AC is used in the same circuit to keep the arc going through the zero value.

Square-wave technology has largely been replaced by inverter-type machines. Inverters have the capability of compressing the AC wave, so that the time between each cycle of AC (frequency) is very short. One wave of HF AC is used in a circuit to keep the arc going through the zero value. Another AC wave stabilizes the arc. With some inverter power sources, the frequency of the AC wave is adjustable. As AC frequency increases, the arc becomes more focused and the molten pool becomes narrower.

AC balance

A balanced AC wave has the same amount of amperage at both electrical poles. This means that the same amount of amperage is output on the straight polarity side (negative pole) and reverse polarity side (positive pole). With AC balance control, the amount of amperage output can be changed to favor one pole over the other.

Each type of polarity has an effect on the welding arc. Straight polarity (-) concentrates the heat in the base metal and allows for deep penetration. Reverse polarity (+) concentrates the heat in the electrode, but the reverse flow of current has a surface-cleaning action. This cleaning action will break up oxides present on the surface of aluminum, magnesium, and copper alloys. See the section on TIG-welding aluminum for more information on surface oxides.

A 90 percent negative and 10 percent positive balance will create a narrow arc with more penetration into the base metal, and little cleaning action on the surface.

A 60 percent negative and 40 percent positive balance will make the welding arc wider and less stable. There will be less penetration into the base metal, but the more time spent at the positive pole, the more cleaning action is created on the surface of the base metal.

EQUIPMENT

Torches

While the business end of a GMAW wirefeed welding machine is called a gun, when using a GTAW welding machine, it's called a torch. GTAW torches are either air-cooled or water-cooled. Air-cooled torches use the flow of shielding gas to cool the torch. Water-cooled torches use a pump and reservoir in a closed system to circulate coolant through the torch.

Typically, water-cooled torches are used with currents above 200 amps. When a short amount of time is spent welding at higher amperages, an air-cooled torch can be used. Usually, air-cooled torches are used with currents less than 200 amps. However, air-cooled torches may not work for long periods of arc on time, even at less than 200 amps. If the heat of the arc makes the torch too hot to handle, either reduce the time spent welding or switch to a water-cooled torch.

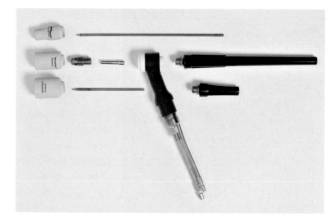

This is a TIG torch with all the accessories disassembled. A variety of torch cups and collets are made to accommodate different-sized electrodes. Short backs are used to gain access into tight places. The long back shown at center right comes standard with most torches. *Monte Swann*

Some air-cooled torches are valved, using a knob to turn the shielding gas on and off. This type of torch is needed when welding with a CC/SMAW machine, which is not equipped with a gas solenoid.

A wide variety of tungsten sizes can be used with every torch. If you plan on using different tungsten sizes, you will need to have the torch parts for each diameter. The torch cup, inner collet, and outer collet (diffuser) are specific to the diameter of tungsten being used.

Cables and hoses

The cables and hoses attached to the TIG torch carry the electrical current, shielding gas, and coolant (for a water-cooled torch). Treat these cables and hoses with care. The tungsten is sharp and remains very hot even after welding has stopped. The slightest touch from a hot electrode or a dropped piece of sharp metal will put a hole in a gas or coolant hose. Fixing a leak in the system will cost both

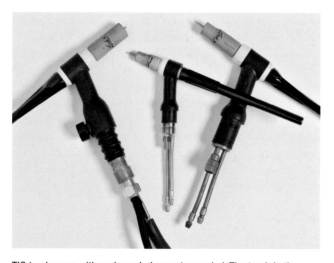

TIG torches are either air-cooled or water-cooled. The torch in the center is water-cooled. Notice it is smaller than the air-cooled torch on the left. The two additional tubes on the water-cooled torches circulate the coolant through the torch. *Monte Swann*

An air-cooled torch with a gas on-off knob is used when scratch starting. Open the valve ¼ turn to start the flow of shielding gas. If the machine is equipped with a gas solenoid, the gas will turn on automatically when the arc is started with high frequency. *Monte Swann*

time and money. Shielding-gas hoses are soft and are easily pinched. If the shielding-gas hose has been stepped on or the wheel of a cart is crushing the hose, you'll immediately notice a change in the arc. The finished weld will have porosity because the shielding gas flow was restricted. Consider using a cable wrap to help protect the hoses and cables.

Cups (nozzle)

The torch cup directs the flow of shielding gas to the molten weld pool. Cups are usually made of a ceramic material (alumina clay) and can break easily if treated improperly. Cup sizes are measured by the inner diameter (ID) and are different depending on the type of torch used. Some manufacturers will assign a number to a cup size (No. 4, No. 6, etc.) which

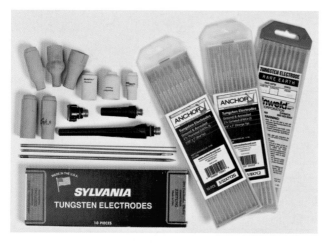

Tungsten is sold in packs of 10. A variety of tungsten alloys are used in TIG welding. Torch cups (pink alumina pieces) and torch backs (black pieces) are specifically designed for use with a particular torch. Set up your torch with the correct-diameter electrode and torch cup for the amperage used. *Monte Swann*

references the actual ID. Typically, a No. 4 cup is 4/16 inch, which equals 1/4 inch. Use the correct cup size for diameter of tungsten; refer to the chart in this chapter for more information. Ideally, the cup diameter should be four to six times the diameter of the tungsten being used. The proper-sized cup will create a column of shielding gas over the weld.

Using a cup too large or too small for a tungsten electrode will alter the shielding-gas stream and flow rate. Small cups will tend to overheat from high amperages used with larger-diameter electrodes or block the flow of shielding gas when a large-sized tungsten is used.

Shielding-gas lenses are sold in kits that fit a variety of torch brands. The lens or screen acts like a filter, providing better shielding-gas coverage over the weld area. Gas lenses evenly distribute the flow of shielding gas and reduce the overall rate of consumption. Since gas lenses are more efficient, less gas is needed and longer tungsten extensions (stickout) can be used. A clear, high-temperature, glass torch cup pushes onto the screen adaptor, allowing a better view of the weld pool.

Inner collet

The inner collet holds the piece of tungsten firmly in the torch, keeping it from slipping out. Slots are machined in the collet so when the torch back is tightened, it clamps down on the electrode. Use the right-size collet for the diameter of tungsten. Poor contact between the collet and electrode will destabilize the arc and overheat the tungsten.

Outer collet (collet body or diffuser)

Shielding gas traveling through the gas hose and torch body comes out the holes of the collet body. Shielding gas is then directed to the molten pool by the torch cup or gas lens. The inner and outer collets are specific to the brand of TIG torch being used and the diameter of the tungsten electrode.

| Electrode Diameter | | Use Gas Cup I.D. | Direct Current, A | | Alternating Current, A | |
in.	mm	in.	Straight Polarity[b] DCEN	Reverse Polarity[b] DCEP	Unbalanced Wave[c]	Balanced Wave[c]
0.010	.025	1/4	up to 15		up to 15	up to 15
0.020	0.50	1/4	5-20		5-15	10-20
0.040	1.00	3/8	15-80		10-60	20-30
1/16	1.6	3/8	70-150	10-20	50-100	30-80
3/32	2.4	1/2	150-250	15-30	100-160	60-130
1/8	3.2	1/2	250-400	25-40	150-210	100-180
5/32	4.0	1/2	400-500	40-55	200-275	160-240
3/16	4.8	5/8	500-750	55-80	250-350	190-300
1/4	6.4	3/4	750-1100	80-125	325-450	325-450

a. All values are based on the use of argon as the shilding gas.

b. Use EWTh-2 electrodes.

c. Use EWP electrodes.

Welding Handbook Volume 2, Eighth Edition, Table 3.2, Reproduced with permission from the American Welding Society (AWS), Miami, FL USA

Torch backs (back cap)

Torch backs insulate the back of the tungsten electrode. They screw onto the torch body and secure the tungsten in the collets when tightened. Torch backs come in different lengths. Short torch backs are used in hard-to-reach places for improved access to the joint being welded. The tungsten will need to be cut or broken in a shorter piece to fit inside the torch back. Standard-sized, long torch backs accommodate a full-length piece of tungsten and are convenient to use. The tungsten is easier to prepare and resharpen when a longer piece is used.

ELECTRODES

Electrodes used for GTAW are pure tungsten or tungsten alloyed with another material. Tungsten is a very hard and brittle metal when properly heat treated, with a very high melting point. The physical property of tungsten allows the material to maintain its shape under the tremendous heat of the arc. It is possible to melt a tungsten electrode. The tungsten will melt when a small diameter is used at a high amperage, or switching the polarity to DCEP. The goal is for the tungsten electrode to maintain its shape during welding. Certain tungsten alloys do a better job of maintaining their shape than others.

Common tungsten sizes are 0.040 inch, 1/16 inch, 3/32 inch and 1/8 inch. The diameter of the tungsten electrode used is determined by metal thickness, amperage, and polarity. Have a variety of size and types of tungsten electrodes available. Remember to have the matching-sized torch parts ready for use.

Tungsten Electrodes for GTAW

Color	Alloy/Name	AWS Classification	Characteristics
Red	Thorium/ 2 percent Thoriated	EWTh-2	• For welding with DCEN polarity • Higher current-carrying capacity • Maintains shape and has a longer life • Easy arc starts and more stable arc • Reduced weld metal contamination • Radioactive
Yellow	Thorium/ 1 percent Thoriated	EWTh-1	• Same characteristics as 2 percent thoriated only slightly less of each • Radioactive
Brown	Zirconium/ Zirconiated or Zirtung	EWZr	• For welding aluminum and magnesium with AC • Higher current-carrying capacity • Maintains shape through AC wave cycle • High arc stability • Reduced weld-metal contamination • Non-radioactive
Green	Pure tungsten; no alloys	EWP	• For welding aluminum and magnesium with AC • Current capacity is lower than zirconiated • Will maintain a ball-shape and a stable arc only at low amperage • Low resistance to weld-metal contamination • Will not maintain a sharp point
Orange	Cerium/ 2 percent Ceriated	EWCe	• For use with either AC or DC • Lower performance compared to thoriated or zirconiated electrodes, but tolerable • Non-radioactive
Gold or black	Lanthana/ Lanthanated	EWLa	• For use with DC • Lower performance compared to ceriated • Does not hold its shape well during arc strikes
Blue or grey	Unspecified oxide(s)	EWG	• Performance depends upon type and quantity of alloying elements • Added elements affect the characteristics of the arc as defined by the manufacturer • Alloys include yttrium oxide, rare earth oxide, and magnesium oxide • Can also include ceriated or lanthanated electrodes with added oxides

GAS TUNGSTEN ARC WELDING (GTAW)

Most filler rods sold today are embossed on the ends identifying the type of filler alloy. Older filler rods had stickers attached to the ends, or no identification on the rod. The filler rods from top to bottom are:

- 1/16-inch ER 70S-2
- 1/16-inch ER 70S-2 with embossed end
- 1/16-inch ER 70S-3
- 3/32-inch RG 60, for gas welding
- 3/32-inch ER 309L with heat number on sticker
- 0.045-inch ER 308L with embossed heat number
- 1/16-inch ER 316L
- 3/32-inch Silicon Bronze
- 3/32-inch Bare Bronze rod, used with a paste flux for torch brazing
- 3/32-inch Pre-fluxed braze welding rod
- 3/32-inch 4043 Aluminum
- 1/4-inch 4040 Aluminum
- Filler rod for welding cast iron
- Gal-Vis rod for coating metals, repair of burned or damaged galvanized coatings, and joining rusty automobile and truck body panels instead of bondo or plastics.

Monte Swann

Tungsten electrodes are color-coded on the ends so that the electrode can easily be identified. Most manufacturers use the same color-code system, but some may not. Consult the AWS classification if you are uncertain about the alloying element in a tungsten electrode. If you decide to sharpen both ends, or break a full rod into smaller pieces, keep track of what type it is so you will not have to guess later. It is almost impossible to tell the different electrodes apart without the color code. The tungsten rod will be easier to break if a small notch is put in it with a file or abrasive wheel. Use two pairs of pliers to make the break. It should come apart at the notch. An empty film can, prescription medicine bottle, or mint tin filled with short pieces of tungsten fits into your pocket. This way you can have multiple sharpened electrodes ready to use while working on your project.

Of all the elements alloyed with tungsten, thorium is the most controversial. Thorium is a radioactive material and exposure to elevated levels is a health risk. When throated tungsten is prepared by grinding, the radioactive thorium dust can be inhaled. There is a risk of internal radiation exposure, especially for people with long careers in welding. There is minimal radiation exposure from storing, handling, and welding with the electrode. If a dust collection system is used when grinding, and the thoriated tungsten is used only with its intended polarity (DC), then the risk factor is greatly reduced. Is it safe to use thoriated tungsten? It is up to the individual to decide, since there is a potential risk involved.

FILLER RODS

Filler rods are used with GTAW to add weld metal to the joint for reinforcement, making it stronger. The filler metal is selected based upon the type of metal being welded, type

Keep a variety of different-diameter filler rods handy. This filler rod rack is a great way to keep filler rods organized and easily accessible. *Monte Swann*

of shield gas used, and the mechanical requirements of the final piece.

Mild-steel filler rods used in oxy/acetylene welding (RG 45, RG 60) are not recommended for GTAW welding since they tend to contaminate the tungsten electrode. If necessary, strips of the base metal can be sheared off and used as a filler metal, but this typically is not recommended.

Filler rods are identified by a set of letters and numbers, which describe the type of alloys contained in the rod.

- ER designation stands for electrode/rod, which means that the same type of filler metal is available for use as a filler rod for GTAW or a welding wire (electrode) for GMAW
- The next two numbers indicate the tensile strength of the filler metal in psi. For example 70 = 70,000 psi tensile strength
- S stands for a solid-wire electrode/rod (with other wires T = tubular or C = metal-core)
- The last digit describes certain characteristics of the filler metal

SHIELDING GAS

Shielding gases are stored at very high pressures in compressed gas cylinders. Be familiar with the hazards related to storing, handling, and using these cylinders. Read the chapter on

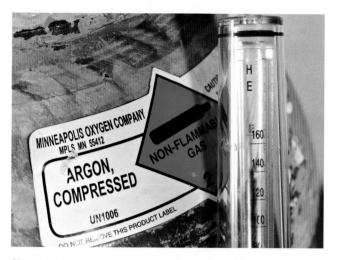

Glass-tube type regulators can sometimes be used for more than one type of shielding gas. Notice argon is being used, but the set of numbers on the regulator tube are for helium (HE). Be certain to read the correct set of numbers for the shielding gas used. *Monte Swann*

Filler rods/Electrodes

Classification/ specification	Metal	Characteristics/ Uses
ER 70S-2	Mild steel	Contains silicon, manganese deoxidizers, along with aluminum, zirconium, tin in small amounts Known as triple-deoxidized wires Properties similar to ER 70S-3 Can weld all grades of carbon steel (killed, semi-killed, rimmed) Can weld dirty, rusty metal; weld strength will vary with the amount of oxides (dirt, rust, scale) on metal Used with CO_2 or Argon-CO_2 in GMAW Preferred wire for vertical and overhead welding using short circuit transfer
ER 70S-3		For welding low- to medium-carbon steels Contains silicon and manganese deoxidizers in greater amounts than ER 70S-2 Produces porous-free welds under normal conditions Fair- to medium-quality welds on rimmed steels (high oxygen content) High-quality welds on killed (fully deoxidized) and semi-killed (low oxygen content) steels In GMAW on rimmed steels: Argon-CO_2 or Argon-O2 shield gas Single-pass welds only Short-circuit, spray, pulse-spray transfers In GMAW on killed or semi-killed steels: Argon-CO_2 or straight CO_2 Single or multi-pass welds Short-circuit, spray, pulse-spray transfers
ER70S-5		Contains aluminum, silicon, manganese deoxidizers For welding killed (fully deoxidized) and semi-killed (low oxygen content) steels only For welding steels with rusty or dirty surfaces Used with CO_2 shielding gas in GMAW Used for spray and pulse-spray transfer only (not used for short circuit) Used in the flat and horizontal positions only (not used in vertical and overhead welding)

Filler rods/Electrodes (continued)

Classification/ specification	Metal	Characteristics/ Uses
ER70S-6		Contains high levels of manganese and silicon for powerful deoxidizing characteristics Can weld over moderate amounts of rust, scale, and dirt on metal surface Reduces the potential for hot cracking on high-carbon steels Used on all grades of steel (rimmed, semi-killed, killed) Used with CO_2 or argon-CO_2 shielding gas High impact resistance when used with straight CO_2 Used in all types of wirefeed transfers Used in all welding positions Single or multi-pass welds Low spatter, good bead appearance on sheet metals
ER 70s-B1	Alloy steels	For welding carbon steels, low-alloy steels, high-strength/low-alloy steels
ER 80S-D2 ER 90S-G		Contains high amount of manganese, silicon deoxidizers 0.50 percent molybdenum added for greater strength For welding steels like 4130 and high-strength T-1; difficult to weld low-carbon steels In GMAW, maximum properties obtained with an Argon-CO_2 shielding gas
ER 308L	Stainless steel	For welding 304, 308, 321, 347
ER 308L-Si		For welding 301, 304
ER 309L		For welding stainless steel to carbon steel and welding 400 series stainless steel
ER 316L		For welding 316
ER 4043	Aluminum	Used to weld the widest variety of aluminum alloys
ER 4643		Used to weld 6000 series aluminum alloys if the metal is to receive post-weld heat treatment
ER 5356		Used for better mechanical properties of 6061 aluminum left in the as-welded condition (no post-weld heat treatment)
ER CuSi	Copper alloys	Silicon bronze
ER Cu		Deoxidized copper
ER Al-(A1,A2,B)		Aluminum bronze

compressed gases for more information on cylinder safety. One of the advantages of TIG welding is that argon is used in virtually all applications. Argon is an inert gas comprised of only one type of atom. Because argon is inert, it doesn't combine with elements in the surrounding atmosphere or with the base metals when they are heated to their melting point (non-reactive). Argon shielding gas reduces the amount of penetration into the base metal, which is an advantage when welding thin-gauge sheet metal or thin-wall tubing. Typical welding grade argon is 99.95 percent (or more) pure and will work for welding most metals.

Higher-purity argon, or argon mixed with small amounts of hydrogen or nitrogen, is used in specialized TIG-welding applications.

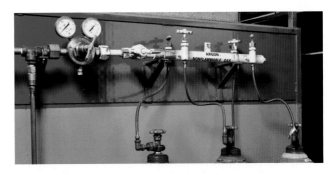

At some shops and schools, a manifold is used to supply shielding gas to multiple welding stations. The regulator reduces the cylinder pressure to 50 psi in the lines. Flow meters at each station regulate the flow of shielding gas used at the torch. *Monte Swann*

Helium is sometimes mixed with argon to create a higher arc temperature at lower amperages. Argon-helium shielding gas mixes will penetrate deeper into the base metal. The mixed gas is used for welding thicker sections of metal, especially aluminum more than ⅜-inch thick or welding with faster travel speeds. Argon-helium mixes are more expensive than straight argon. Pure helium was the first shielding gas used for GTAW, but is not commonly used today. Straight helium is more expensive than argon-helium mixes, produces deeper weld penetration, and requires two to three times more volume (than argon) for adequate shielding.

Do not use C25 or other gas mixes used for wirefeed welding. Base metals welded with GTAW will become contaminated if exposed to CO_2 or oxygen.

PURGE GASES

When TIG welding a joint with an open root, the backside of the weld can become contaminated by the atmosphere. Since shielding gas from the torch only protects the face side of the weld, a backup purge gas is used to protect the backside. There are several techniques for using a purge gas. An argon or helium purge gas is used on certain types of metals, such as stainless steel and titanium, in specific applications. Purge gases should flow at about 6 CFH during welding operations. Simple jigs, cardboard, and tape can be used to make an enclosure to block off the backside of the weld joint from the surrounding air. A separate flow meter is used with hoses at each end of the enclosure for gas flow in and out. Paste products, such as solar flux, are used as an alternative to a purge gas. The paste is applied to protect the backside of welds, usually with mixed results. Flux-coated or flux-cored filler rods are also used to weld unpurged pipes.

REGULATORS/FLOW METERS/FLOW RATES

Regulators are specifically designed for use with certain gases. Most regulators for argon will also handle argon-helium mixes. A regulator changes the flow of gas from the cylinder (at a high pressure) to the hoses (at a much lower pressure). Often, a flow meter is combined with the regulator so the rate of shielding gas flowing from the torch cup can be adjusted. Shielding-gas flow rates are measured in cubic feet per hour (CFH). Because argon is heavier than air, low flow rates can be used. For GTAW with straight argon, a flow rate of 15 to 20 CFH is usually adequate. For larger-diameter electrodes, longer electrode stickouts and higher amperages, the rate can be increased up to 35 CFH. Because helium is lighter than air, a higher flow rate is required. For GTAW with Helium mixes, the flow rate will need to be increased to 30–50 CFH, depending on the amount of helium in the mix.

If the shielding-gas flow rate is too low, it will not provide enough coverage over the molten puddle. This is especially true when longer tungsten stickouts are used. If the flow rate is set too high, turbulence from the gas stream can aspirate air, sucking in oxygen, nitrogen, and hydrogen, and

contaminating the weld.

ADVANTAGES

By using GTAW, the widest variety of metals can be welded with the least amount of setup time and expense. With an AC/DC GTAW machine, all it takes to switch from welding steel to welding aluminum is changing a few machine controls, using a different filler rod, and changing the electrode. In some cases, even the same electrode can be used for both steel and aluminum.

Very small welds can be made with pinpoint accuracy.

A regulator/flow meter is combined for use with individual welding machines. The gauge indicates the amount of gas left in the cylinder—in this case it is full at 2,600 psi. Open the cylinder valve slowly to reduce the impact of the high-pressure gas rushing into the regulator. *Monte Swann*

Turn the adjustment knob on the flow meter until the top of the ball is at the required CFH. In this case, the ball reads 35 CFH, which is too high a flow rate for most GTAW applications. *Monte Swann*

Use a tight-fitting wrench to check the connections in the gas lines. A loose connection can suck air into the lines, mixing with the shielding gas and contaminating the weld. Notice this high-pressure cylinder has a safety nut and disc on the right side of the valve stem. *Monte Swann*

In pipe welding, the root pass can be made with gas tungsten arc welding. The root pass on the left side, between the 12-inch diameter pipe and flange, was made using the GTAW process. The root pass on the right was made with an E-6010 SMAW electrode. Each joint will be finished with E-7018. *Monte Swann*

The GTAW process works well for joining a variety of metal thicknesses. It was easy to weld the thick-walled tubing to a piece of 14-gauge sheet metal. *Monte Swann*

The TIG process can also be used for brazing. A piece of mild steel and stainless steel are joined with a silicon bronze filler alloy. The metals are heated to brazing temperature (below the melting point) with the TIG arc, then filler alloy is applied to the joint. Silicon bronze has a high-strength, similar to carbon steel. *Monte Swann*

There is more control over the heat input into the base metal and the amount of filler metal added to the joint than in the other manual welding processes. Welds can also be made without additional filler metal (autogenous weld).

DISADVANTAGES

Because a shielding gas is used to protect the molten weld pool, GTAW will be disrupted by any air movement over 5 miles per hour, which blows the shield gas away. This instantly contaminates the weld area causing porosity in the bead. GTAW is not typically used outdoors, but putting up screens or a tent made from a non-flammable material can help solve the problem. Torches outfitted with gas lenses are more wind resistant, and there are little shields that attach to the torch to help direct the flow of gas.

Since the filler metal has to be manually fed into the molten puddle, it takes a much longer time to build up weld bead. Travel speeds are slower. Metals ½-inch thick or more are usually welded with another process. The process also requires a higher level of skill and more time and practice to use effectively.

PREPARING TUNGSTEN

Tungsten electrodes come unsharpened. Before TIG welding, tungsten more than ⅟₁₆ inch in diameter should be ground to a point. Tungsten ⅟₁₆-inch diameter and smaller can be used with a broken end, they do not need to be ground. There are several ways to grind a piece of tungsten:

- Use a machine specifically for grinding tungsten. Tungsten grinders come in all different sizes and use a diamond wheel to grind the metal. The tungsten is inserted into the machine and held at specific (adjustable) angles.
- A pedestal grinder with a stone wheel
- A belt sander
- A chemical sharpener, which works by dipping hot tungsten in the chemical solution multiple times. In my experience, chemical sharpeners do not perform as well.

If a pedestal grinder or belt sander is used, a belt or stone wheel should be dedicated to grinding tungsten only. If other metals, such as steel or aluminum, are ground with the same abrasive surface, those metals become imbedded in the electrode. When the contaminated tungsten is used for welding, the metal particles can transfer into the molten weld pool, contaminating the weld.

Always sharpen tungsten with it pointed down. It is possible to sharpen tungsten pointed up, but if the piece gets caught on the belt or wheel, it will be shot down at a high speed. If this happens, the tungsten rod can end up going through your hand.

If you have a portable drill handy, the tungsten can be

Any of the tungsten alloys (indicated by the color band on the end) can be prepared with this small tungsten grinder. Larger-sized grinders are available for more frequent grinding. *Monte Swann*

put in the chuck and turned at a low speed. Otherwise, the tungsten will need to be rotated by hand in order to create the desired cone shape at the end. A small pin vise can be used for a better grip on the tungsten. This is especially helpful for turning and sharpening shorter pieces.

Use light to medium pressure when sharpening. If too much pressure is used, the tungsten will heat up from the friction and turn blue. A color change indicates changes to the mechanical properties of the electrode, reducing its performance.

Once a piece of tungsten is sharp, keeping it that way is a challenge. Anytime the electrode is touched to the work, both the base metal and tungsten become contaminated. The tungsten should be resharpened when this happens. If there is a risk of tungsten inclusions in the weld, the base metal may need to be ground out as well. If the end of the tungsten is completely messed up from base metal contamination, use two pliers to break off the end, then resharpen.

It is important to know that electricity flows on the outside of the tungsten electrode. For this reason, the grinding marks should always be in line with the length of the tungsten (lengthwise). If you sharpen a piece of tungsten perpendicular to an abrasive wheel or belt, the grinding marks will also be perpendicular. The electricity will follow the grinding grooves in the tungsten. Perpendicular (circular) grooves will spin the arc coming off the end, making it less stable and less focused.

Tapers

When a piece of tungsten is sharpened, the amount of taper will determine the width of the arc. With a wide taper (greater included angle), the electricity flows down the tungsten, turns at the taper, and comes off the end at a similar angle. This will make the arc spread out. With a shallow taper (lesser included angle), the arc coming off the end will be narrow and focused. You may use different tapers, depending on the joint design, thickness of metal, and amperage. Typically, taper length is

2–3 times the diameter of the tungsten. Use a very sharp, smooth taper (needle point) when welding stainless steel for a stable, focused arc.

Balling tungsten

Traditionally, the sharp, tapered end of the electrode was balled when using AC. Balling is accomplished by switching the machine to DCEP, positioning the tungsten over a piece scrap metal of the same type to be welded (aluminum). Starting an arc for just a few seconds causes the tip to melt into a ball shape. When using transformer-rectifier machines, this is still a good way to prepare the tungsten for welding with AC. Inverter machines use HF AC, making it unnecessary to pre-ball the tungsten. Either the action of the current will ball the end slightly or not change the shape at all. To ball or not to ball depends on what type of machine you are using. If the ball becomes large from too much current, switch to a larger-diameter tungsten or a different tungsten alloy.

Prolonging electrode life

Use the proper current for each tungsten diameter to prolong electrode life. Electrodes are commonly used with amperages in the high range (high current density) for maximum arc stability. Using amperages in the middle of the current range reduces electrode contamination when accidentally touched to the base metal. Short tungsten stickouts and proper gas-flow rates will protect the hot electrode from the atmosphere. Resharpen the tungsten if it becomes contaminated. Keep an even, cone-shaped taper on the end.

Tungsten stickout (electrode extension)

The distance from the torch cup to the end of the tungsten is called electrode extension or tungsten stickout. The problems with a short stickout are the torch cup may interfere with your view of the molten weld pool or prevent access when trying to weld joints with tight clearances. If the stickout is too long, the shielding gas protecting the weld pool and tungsten electrode may thin out. A long column of shielding gas can easily be blown away. Use different electrode extensions depending on the joint design and torch angle. Torches retrofitted with gas lenses can have a longer tungsten stickout. A clear torch cup will not interfere with the view of the weld pool.

SETTING AMPERAGE

There are many factors to consider when setting the amperage for GTAW. Welding ⅛-inch aluminum with AC will require more amperage than welding the same thickness of mild steel with DCEN. When using a remote amperage control, the setting on the machine should be the maximum value. If I'm using a foot pedal control and I set the machine for 120 amps, when I start the arc, the amperage is much lower than 120. When the pedal is pushed all the way down, the output is 120 amps. The foot control is like the gas pedal on a car. The more you put your foot down, the higher the amperage and

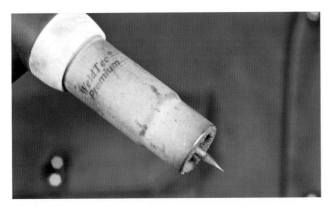

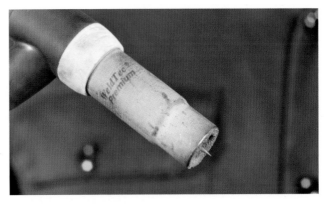

When working with tungsten electrodes, you want about this much of it protruding, giving you a longer stickout. Longer tungsten stickout might be required to maintain the proper distance between the electrode and the workpiece in joints that are more difficult to access, like tee joints. Using a shielding gas lens on the torch will allow for longer tungsten stickouts than with a traditional ceramic cup and diffuser. *Monte Swann*

A very short tungsten stickout will make it difficult to see the molten pool blocked by the torch cup. *Monte Swann*

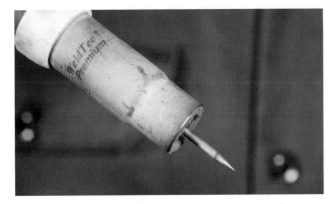

Too much tungsten stickout may compromise shielding of weld pool. *Monte Swann*

In some conditions, you will want your tungsten stickout shorter. This is the proper amount of tungsten stickout for those conditions. A shorter tungsten stickout can be used for butt joints where access to the joint is not restricted. A shorter stickout will provide better shielding of the molten weld pool. *Monte Swann*

the hotter the arc. If you are a lead foot, set the amperage on the low side to start so you will not burn through from too much heat. I prefer to have enough amperage to create the molten puddle before the pedal is halfway down. Setting the amperage this way is especially helpful when using a torch-mounted switch. The control is more difficult to move, so if the amperage is set on the high side, less time is needed to get the base metal hot enough to weld.

WELDING TECHNIQUE

The TIG welding torch is pushed. Travel in the direction your torch is pointed (forehand technique).

GTAW welding is a little more difficult than the other welding processes, because several skills are used simultaneously. Think about someone learning to drive a car. At first, it's difficult concentrating on pushing the gas pedal and steering the car while paying attention to the traffic on the road. With time and practice, steering becomes automatic. Turning corners and switching lanes is easier. The foot on the gas pedal develops a mind of its own. Instead of thinking about letting off the gas to slow down, your leg lifts up automatically when traffic slows down in front of you.

TIG welding gets easier with practice, as well. Eventually your hands will develop a kind of intelligence, so your mind can concentrate on what is happening with the weld pool. With practice you will be able to make adjustments without thinking too much about it. TIG welding, or any welding process, may feel awkward at first—keep practicing. In time, your mind and body will be in sync and welding will become more instinctive and less difficult.

In gas tungsten arc welding, one hand holds the torch, the other hand holds a filler rod, and, in many cases, a remote amperage-foot-control switch is used. Start off only using the torch and foot pedal. Add the filler rod to the equation when your foot controlling the amperage is coordinated with your hand manipulating the torch. The techniques for pushing a puddle and adding filler rod are similar for gas welding and GTAW.

No matter if you are a beginner or expert, everybody will

	Metal thickness	Polarity	Amperage	Filler rod diameter
Mild steel and low-alloy steel	24 gauge (.035)	DCEN	20–80	1/16
	18 gauge (.050)		80–100	
	1/16 (.063)		80–120	
	3/32 (.093)		100–140	3/32
	1/8 (.125)		100–140	3/32
	1/4 (.250)		130–175	1/8
	3/8 (.375)		170–200	3/16
Stainless steel	24 gauge	DCEN	20–50	1/16
	18 gauge		50–80	
	1/16 inch		65–100	
	3/32 inch		85–130	
	1/8		110–150	3/32
	1/4		200–350	3/16
Aluminum	1/16	AC	60–90	1/16
	1/8		115–135	3/32
	3/16		190–240	1/8
	1/4		220–260	1/8, 3/16
Grey cast iron	1/4	DCEN or AC	150–160	3/16
	1		300–350	3/16 for first pass, larger sizes for others

occasionally touch their tungsten to the base metal. When it happens, either resharpen the tungsten or grab another sharpened piece ready to go. Trying to weld with a contaminated electrode will make welding difficult. Resist the temptation to just keep welding if your tungsten is messed up.

Scratch start

The TIG arc can be started by scratching the tungsten electrode against the base metal, and then bringing it up slightly to begin welding. This process is similar to striking an arc with a stick welding electrode. Any CC machine can be used to scratch start the arc. When machines designed for SMAW are used for GTAW, the arc will have to be started this way. Only DC polarity is used with a scratch start.

Starting a bead

If your machine has an HF start, bring the torch close to the base metal at the beginning of the joint. Then use the remote amperage control to start the arc. When using lift or scratch starts, strike the arc in the joint to be welded or use a separate piece of aluminum, copper, or steel as a start block. Once the arc is started, it can be carried over to the work piece. As soon as the arc is struck, lift the torch 1/8 inch above the base metal to avoid contaminating the electrode in the molten puddle.

Once the arc is started, the heat will begin to form a molten puddle. On thicker metals, it may take a few seconds for the base metal to heat up enough for the metal to melt.

If you have to wait a few seconds, the heat of the arc has a preheating effect. This preheat slows the cooling rate of the base metal. Preheat is not a problem with most metals. Slowing the cooling rate with preheat is required for metals hardenable by heat treatment to prevent cracking. However, slowing the cooling rate is not advisable when welding stainless steel.

On thinner metal, start with a low amperage and slowly increase it until the base metal melts. Blasting the metal with a surge of amps can thermally shock the metal from the rapid change in temperature. Too much heat will cause excessive penetration, burning a hole through the base metal.

The same technique is used for adding filler metal in TIG welding and gas welding. Add filler rod to the molten pool by dipping it in the center or leading edge of the puddle. Do not let the filler metal melt off in droplets and fall into the puddle.

Ending a bead

When finishing a bead, lower the amperage slowly. Be certain to fill the crater at the end of the weld with enough filler rod. A concave crater, especially on aluminum, is more prone to cracking. When the arc has stopped, don't pull the torch away. Wait at the end of the weld for a couple of seconds so the shielding gas can protect the solidifying filler metal and tungsten electrode.

Riding the cup (walking the cup)

Since the torch cup is made of a non-conductive material, it can be in contact with the base metal while welding. This TIG welding technique is useful in awkward situations where the tungsten electrode has to be supported at a specific angle and distance. Resting the cup on the base metal will work well for getting the tungsten very close and keeping it stable when making small tacks on thin pieces. Riding the cup will not work for everything. Learn to TIG weld without using it, so that it remains an asset rather than a crutch.

Porosity

There are three main causes for porosity in the weld metal:

1. The base metal can contain a large amount of iron oxides. When the molten puddle is formed, the iron oxides combine with carbon in the metal creating carbon monoxide/dioxide. When released, these gases cause porosity. A filler rod with deoxidizing agents can help reduce the formation of carbon monoxide/dioxide gases.

2. Rust or scale on the surface of the metal also contains iron oxides. Dirt, oil, grease, or moisture on the filler rod or base metal vaporize when heated. This vapor can mix into the weld pool, trapping gases and causing porosity.

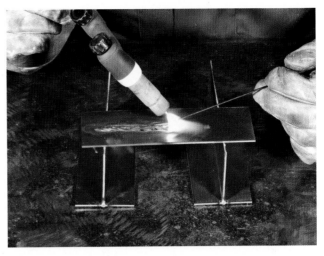

If the tungsten electrode is too far from the base metal, the arc spreads out, heating a large area. *Monte Swann*

3. A lack of shielding-gas coverage leaves the molten weld pool open to contamination from the atmosphere. Inadequate shielding-gas flow is caused by; a gas flow set too low, arc length too long, too much electrode stickout or air drafts.

DASH

The relationship between distance, angle, speed, and heat (DASH) is critical in all welding processes—especially GTAW. If these four factors are managed correctly while running a bead, it will turn out right. Because each factor has an effect on the other, there is no hope of success if attention is not paid to all four variables. DASH changes depending on the welding process used, thickness of metal, joint design, and welding position. Keeping the DASH relationships in mind increases your chances of making a great weld.

Distance

Keep the tip of your electrode as close to the base metal as possible without touching it. The optimum arc length is about $\frac{3}{32}$ to $\frac{1}{16}$ inch, which is very short. Distance (arc length) should remain consistent during the entire weld.

Relationship to *heat*:

Raising the TIG torch will cause the arc to spread out, making a wider molten pool. A shorter arc length will focus the arc, making the weld pool small. Do not change the distance to change the amount of heat. Only change distance to increase or decrease the size of the molten puddle.

Angle

There are two angles to consider. Travel angle is the angle between the torch and the weld axis. Work angle is the angle of the torch perpendicular to the weld axis. Both angles will change depending on the joint design and welding position.

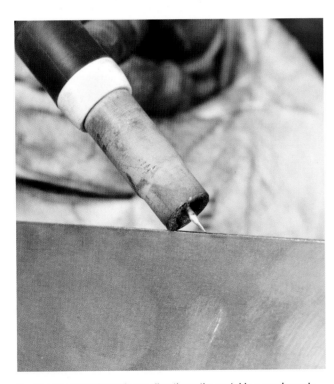

Resting the ceramic torch cup directly on the metal is a good way to steady the torch when tacking. If the torch is moved along the joint while welding, it is called walking the cup. Be careful not to touch the electrode to the base metal. *Monte Swann*

The direction of travel should be a forehand technique, pushing the torch. Each exercise will give recommended angles. Make adjustments to the torch angle to manage and direct the molten puddle.

Relationship to *heat*:

The heat input in the base metal will change depending on your torch angle. Steep angles direct the heat of the arc toward the work piece and increase the amount of penetration into the base metal. If you are burning through the backside of thinner materials, try decreasing the angle of your torch. Shallow angles (less than 35 degrees from the base metal) are not recommended since the arc spreads out and less heat is applied to the base metal. When steep torch angles are used, the rod tends to get drawn onto the hot tungsten instead of the molten puddle.

Speed

A consistent travel speed will put a uniform amount of heat into the base metal along the weld joint. Maintaining a consistent travel speed is the key to creating uniform weld beads. When adding filler metal to the weld pool, dip the end of the rod into the molten puddle frequently and at a consistent rate. When joining thinner sheet metals, move the torch in a straight line. Use a circular or zigzag motion when joining thicker metals.

Relationship to *angle*:

Using the correct work and travel angles will direct the arc into the root of the joint. A consistent weld-puddle size makes it possible to maintain a uniform travel speed.

An incorrect torch angle will direct the heat of the arc to one side. This becomes a critical factor when two pieces of metal are joined. Use correct torch angles for each type of joint and welding position. *Monte Swann*

When the amps are set too high, the base metal overheats and the weld pool spreads out. This may cause melt-through even if filler metal is added. *Monte Swann*

When the amperage is set too low, there is not enough heat for proper fusion. The bead is excessively convex due to the filler metal building up on the surface and not spreading out. There should be a smooth transition between the bead and base metal at the toes of the weld. *Monte Swann*

Relationship to *heat*:

When the same amperage setting is used, a faster travel speed will have less heat input than a slower travel speed. However, the amount of amperage has a large impact on travel speed. A faster travel speed is possible when higher amperages are used. This can put less heat into the base metal. Lower amperages require a slow travel speed in order to maintain the molten puddle. This heats up the base metal for a longer period of time.

Heat

In GTAW, the amount of heat is determined by the amount of amperage. The thickness of the base metal determines the amperage setting. Tungsten electrodes are designed to work within an amperage range. If large-diameter electrodes are properly prepared (grinding a sharp, long tapered point), they can be used successfully at low amperages. Small electrodes will not hold up to the heat and arc force when used at amperage settings higher than the recommended range.

Heat is also controlled by the addition of filler rod into the molten puddle. As the cooler filler metal is added, it reduces the heat in the weld pool. Frequently dipping the filler rod into the puddle helps prevent the liquid metal from spreading out or melting through the base metal. The only way to build up a consistent weld bead is to add filler metal during the entire weld.

Relationship to *angle*:

Torch angle will determine where the heat is placed. Incorrect torch angles will direct heat to where it is not needed, for example, a thinner section of metal or one side of a weld joint and not the other.

Relationship to *speed*:

High amperages require faster travel speeds. Reduce the amount of amperage to gain more control over the molten puddle.

SETUP

Base metals used in TIG welding must be clean. Any contaminants on the surface of the joint, such as rust, paint, dirt, oil, mill scale, and heavy oxide layers, must be removed at least 2 inches from each side of the joint. Next, select the polarity, electrode, and filler rod best suited for your base metal. You may benefit from doing a little research to find out what will work for your project. Taking the time to look up the information may save time, money, and big headaches later.

Learn all the controls on your machine and what they are used for. Start out with the basic controls. Use the features to your advantage, but try not to make things overly complicated when you start out. Check that your shielding-gas flow is set correctly and make sure there are no leaks in the system. If you can't get rid of porosity in your bead, check to see that air is not being aspirated into the system from a lose connection or damaged hose.

TIG welding with the incorrect type or size of tungsten, torch cup, and filler rod is like trying to win the Daytona 500 with four flat tires. Not matter how good a driver is, without the car running right, there is no chance of finishing the race.

WELDING EXERCISES

I recommend completing these exercises in the order presented. The level of difficulty increases with each exercise, so become proficient with your new skills in one exercise before moving to the next. Wear the proper PPE (refer to chapter 2 for more information).

For the GTAW welding exercises you will need:

CC welding machine

3/32-inch-diameter tungsten electrode (I have chosen to use 2 percent thoriated)

ER 70S-3 filler rods, 1/16-inch diameter

Argon shielding gas

14-gauge, cold-rolled, mild-steel pieces cut 2 × 6 inches

Note: All exercises are to be performed with the forehand technique (push). The torch is moved in a straight line; no circular or zigzag motions are used.

EXERCISE 1

Full-open corner joint, no filler metal (autogenous weld)

Torch angles:

Work angle: 90 degrees

Travel angle: 60 degrees

Tack the two pieces of metal together at a 45-degree angle, lining up the tips of the corners. The two pieces should not overlap each other. If one end of the joint is lined up correctly, it can be tacked first. The untacked end can be moved around. Use pliers to hold the metal and tack the second end so there are no gaps in the joint; use a low-amperage setting. Practice this exercise several times to get comfortable moving the torch and operating the foot pedal at the same time. Observe how the amount of heat varies using the remote amperage control. Use two hands on the torch and brace your arms or wrists.

Get setup to make a great weld. The work clamp is attached to the table where some of the paint has been sanded off for a good connection. Both hands are on the torch and elbows rested on the table to comfortably hold the torch and keep it steady for a long period of time. *Monte Swann*

Ease onto the foot pedal until the two pieces fuse together.
Monte Swann

Push the puddle down the joint, maintaining enough amperage to melt both pieces. Keep a close distance the entire length of the joint.
Monte Swann

The finished corner should have rounded edges and complete fusion. Look for any spots along the joint that were not fused together. With GTAW, areas of incomplete fusion can be fixed by remelting the base metal. *Monte Swann*

EXERCISE 2
Bead on a plate with filler rod
Torch angles:

    Work angle: 90 degrees

    Travel angle: 45–55 degrees

    Filler rod angle: 15–25 degrees

An important basic skill in TIG welding is pushing the molten puddle and adding filler metal to it at the same time. Start by pushing the puddle; get used to how fast you need to travel for the amperage used. Start with a lower amperage and slow travel speed. It is easier to learn how to feed the filler rod into the weld pool when moving slowly. As you establish a rhythm of moving the torch and dipping the filler rod, slowly increase the amperage and travel speed. Be careful to add filler metal to the puddle only, keeping it clear of the tungsten. It is easy to touch the filler rod to the tungsten by mistake, especially if the molten pool on the base metal is too small. When this happens, the filler rod fuses onto the electrode instead of the base metal. The electrode will become contaminated and have to be reground.

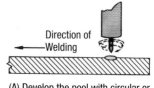

(A) Develop the pool with circular or side-to-side motion

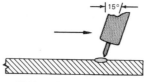

(B) Move electrode to trailing edge of pool

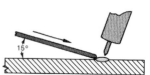

(C) Add filler metal to center of leading edge of pool

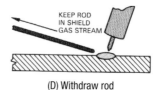

(D) Withdraw rod

(E) Move electrode to leading edge of pool

Welding Handbook Volume 2, Eighth Edition, Figure 3.24, Reproduced with permission from the American Welding Society (AWS), Miami, FL USA

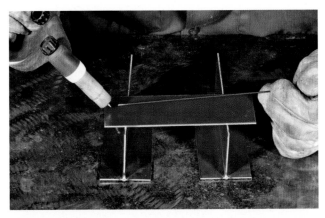

Begin with the tungsten close to the base metal. Keep the filler metal nearby and ready to be added when the base metal melts. Don't pull the filler rod more than ¼ inch away from the molten puddle. That way filler metal stays in the argon shielding gas and can be added to the weld pool frequently. *Monte Swann*

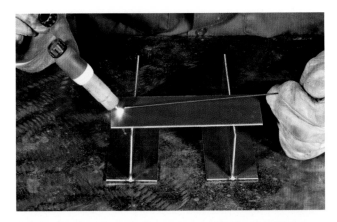

Start the arc; slowly increase the amperage until a liquid puddle forms on the base metal. Begin adding filler metal to the puddle. *Monte Swann*

A circular motion can be used, adding filler metal when the torch is at the trailing edge of the pool. Otherwise, filler rod can be added to the puddle as it is moved in a straight motion across the plate. A slightly higher amperage and travel speed are used with a straight-line welding technique. *Monte Swann*

Add the filler rod by dipping it into the molten puddle. Do not melt off droplets into the puddle or lay the rod down into the joint. If the filler rod sticks to the base metal, increase the amperage and move the torch forward. Make sure to blend any filler metal back into the base metal. *Monte Swann*

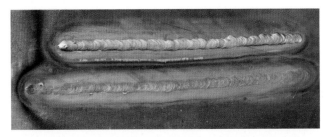

The top bead was made with filler rod. The bottom bead was made without filler rod. Either way, the bead should be a uniform width with even ripples. *Monte Swann*

EXERCISE 3
Square groove butt joint, open root, flat position
Torch angles:

> Work angle: 90 degrees
> Travel angle: 45–55 degrees
> Filler rod angle: 25–35 degrees

Tack two ends of joint together with a ¹⁄₁₆-inch root opening by melting the corners with the torch and adding filler metal to bridge the gap. Keep the amount of root opening consistent along the entire length. A back-stepping technique was used in this exercise to reduce the amount of distortion in the base metal. With back-stepping, the heat from welding is more evenly distributed in the base metal. Using a back-stepping technique is optional. The joint can be welded from one end to the other in a continuous bead. When a continuous bead is made, more heat is concentrated at the end of the weld. Keep the tungsten electrode and the welding arc centered over the root opening. Use enough amperage to melt both sides of the joint and add filler. If only one side is melting, check the work angle; the torch may be tipped to one side. If the arc blows through the joint, reduce the amount of amperage and/or maintain a closer distance with the electrode. When the tip of the tungsten is too far away, the arc will spread out, overheating the edges.

Rest both arms on the table in order to hold the torch and filler rod steady. Precise, controlled movements are necessary for successful TIG welding. *Monte Swann*

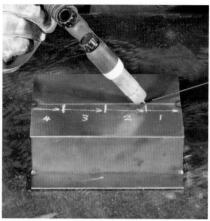

The start and stop points were marked with soapstone. Numbers indicate the order of welding; arrows indicate travel direction. *Monte Swann*

Begin with the first bead and travel to the end of the joint. At the end, wait a few seconds, holding the torch over the solidifying metal to protect it from the atmosphere. *Monte Swann*

Restart the arc and travel to where you began the first bead. Add enough filler metal at the end and fuse the two beads together. *Monte Swann*

Repeat this process for the rest of the beads. Notice the direction of travel is different than the overall progression. This aspect is unique to the back-stepping technique. *Monte Swann*

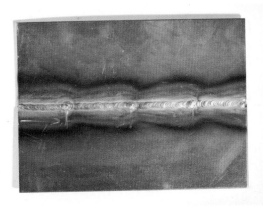
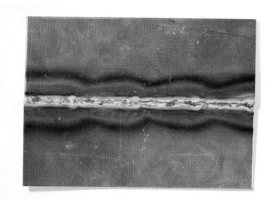

Look at the heat-affected zone and temper colors in the finished weld. You can tell how this joint was welded by the shape of the heat marks. Notice they are narrow where the beads were started, and wider at the end. There is penetration through the joint, although some areas of the root are slightly concave. *Monte Swann*

EXERCISE 4
Lap joint, flat position
Torch angles:

      Work angle: 45 degrees from the base plate (upper plate)

      Travel angle: 45–55 degrees

      Filler rod angle: 25–35 degrees

Tack the ends of the two pieces so half their widths overlap. The joint is at a 45-degree angle for a flat position fillet weld. Concentrate the arc on the base plate. Stay close enough to melt the edge, adding filler metal into the puddle on both sides of the joint. It is easy to melt away the edge, or not melt it at all. Keep the torch steady; don't move the arc away from the joint or change the work angle too much. Position your head so you have a clear view of the weld, moving the torch and weld bead toward you.

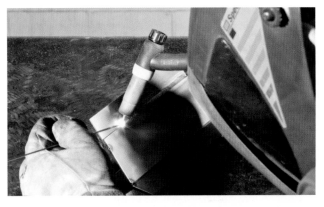

Move along the joint, adding filler metal to the weld pool. If the edge begins to melt away, reduce the arc length by holding the torch closer, or tilt the torch to direct the arc back to the base plate. You may see a notch-shape form in the weld bead. Fill in this notch—the entire length of the seam—for complete fusion at the root of the joint. *Monte Swann*

To begin the fillet weld, tilt the torch so the arc is directed more to the base plate. After the molten pool forms, tilt the torch back to 45 degrees to melt in the edge. *Monte Swann*

The finished bead should be flush with the top piece and have good fusion into the base plate. *Monte Swann*

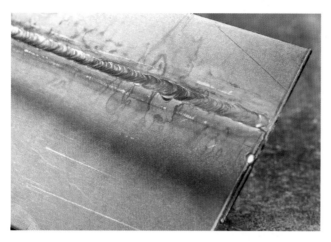

One small movement with the torch momentarily changed where the arc was directed. That slight movement melted part of the edge away. *Monte Swann*

An increased tungsten stickout allows the tip of the electrode to be in close proximity to the root of this tee joint. *Monte Swann*

EXERCISE 5

Tee joint, flat position

Torch angles:

> Work angle: 45 degrees from each plate
>
> Travel angle: 45–55 degrees
>
> Filler rod angle: 25–35 degrees

Adjust your tungsten stickout so the electrode is close enough to the root of the joint. In a tee joint, the torch cup may be closer to the edges of the metal. If not enough stickout is used, the tungsten may be too far away and the arc length too long. A long arc length will melt into the sides of the plate, but not to the bottom of the joint (root). Higher amperages are required for a tee joint. Increase the amount of heat until both pieces of metal melt and the root has completely fused. If only one side of the joint melts, tilt your torch toward the unmelted side; add filler metal to the molten pool.

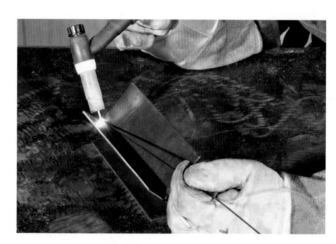

Ease on the foot pedal and increase the amperage until a weld pool forms on both pieces. *Monte Swann*

Hold the torch handle at the end—near the cable and hoses—instead of near the electrode. This grip keeps your hand further away from the heat of the arc. *Monte Swann*

Add filler metal to the middle or leading edge of the weld pool. If filler metal is added before the weld pool forms, there will be incomplete fusion at the root of the weld. Inadequate current and fast travel speeds can also cause incomplete fusion. *Monte Swann*

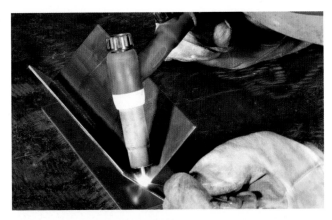

At the end of the joint, back off the foot pedal to reduce the amperage. Tilt the torch, increasing the travel angle to 60 degrees. This keeps the arc over the joint so it does not blow off the edge. *Monte Swann*

The finished weld should have equal legs and good fusion into the base metal without undercut. Look for a uniform width and even distribution of filler metal in the joint. *Monte Swann*

EXERCISE 6
Square groove butt joint with 1/16-inch open root, horizontal position
Torch angles:
    Work angle: 80–85 degrees
    Travel angle: 45 degrees
    Filler rod angle: 25–35 degrees

EXERCISE 7
Square groove butt joint with 1/16-inch open root, vertical position, traveling uphill
Torch angles:
    Work angle: 90 degrees
    Travel angle: 45–55 degrees
    Filler rod angle: 25–35 degrees

Groove welds in the horizontal and vertical positions are made with the weld metal perpendicular (90 degrees) to the work table. The difference is the position of the weld axis (seam)—horizontally or vertically aligned. Place the joint at the edge of the table so the torch can be held at a comfortable angle. Rest your wrists on the table, support your arm on your leg, and get as comfortable as possible. Steady your arms by keeping your elbows tucked in close to your body and use a loose grip on the torch. The technique for horizontal- and vertical-position groove welds is the same one used in the flat position. The challenge is keeping the torch and filler rod steady enough to make small controlled movements. Since the GTAW weld pool is so small, the surface tension of the liquid metal is enough to counteract the effects of gravity.

Position the joint so the filler rod hand is braced against a solid object. This way you can consistently feed filler metal into the molten pool with more accuracy and less fatigue. *Monte Swann*

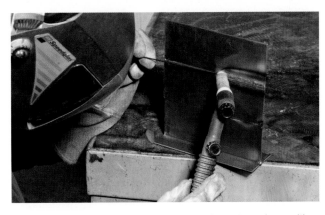

Travel along the joint, centering the arc over the root opening, melting both edges and adding filler metal. *Monte Swann*

Whether traveling upward or downward, the torch is pointed up and filler rod is fed from the top. If you travel downward, use a backhand (pulling) technique. To compensate, change the travel angle to 80–90 degrees. *Monte Swann*

When adding filler rod from the top, it is very easy accidentally to touch it to the tungsten. This will fuse the rod onto the electrode. Solve the problem by maintaining enough heat in the base metal for a weld pool to form and keep the filler-rod hand steady. *Monte Swann*

Decrease the amperage at the end. Keep the torch and shielding gas over the solidifying metal for a few seconds after the arc is off. Avoid pulling away the torch too soon. *Monte Swann*

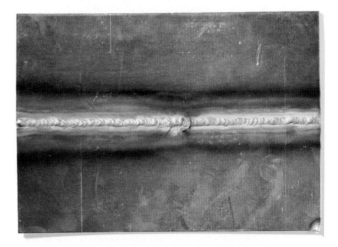

The right half of the weld was made in the horizontal position. The rest was finished traveling vertical upward. The goal is for the finished bead to look like it was welded in the flat position. *Monte Swann*

CHAPTER 13
CUTTING PROCESSES

In the chapter on general tools and equipment we looked at several different ways to cut metal. Band saws, metal-cutting circular saws, chop saws, fiber cutoff wheels, and nibblers are mechanical methods for cutting metal. In this chapter, we will look at three non-mechanical processes for cutting metal. The first process, torch cutting, utilizes the same equipment used for gas welding and brazing, except with a different torch attachment. Plasma cutting requires a machine specifically designed for the process. Carbon-arc cutting can be performed with an SMAW power source and compressed air.

SAFETY IN ALL CUTTING PROCESSES

Read the chapter on general safety before using any cutting equipment for the first time. Read and follow all the safety materials and owner's manuals that come with your machines and equipment. Knowing the hazards involved will greatly reduce the risk of property damage and personal injury. Fires are likely to occur because proper precautions were not taken before cutting, such as not wearing the proper footwear. Remember that sparks can travel up to 35 feet, turn corners, and can disappear out of sight. Sparks are especially aggressive when cutting in the horizontal and vertical positions. Contain and control the ignition source and limit the amount of combustibles in the area you are working.

1. Suit up before cutting; wear the proper PPE for the job.
2. Keep a large-sized fire extinguisher around at all times.

Like oxy/acetylene torches, plasma cutters can be used to make horizontal, vertical, and overhead cuts. *Monte Swann*

3. Remove all flammable materials from the area and at least 40 feet away from any cutting operation. This includes batteries.
4. If cutting is done near flammable materials, use sheet metal partitions to deflect sparks and hot metal.
5. If cutting is done over a wooden floor, sweep the floor clean and wet it down before cutting. Use a metal bucket filled with water or sand to catch the dripping slag.
6. Whenever possible, cut metals in a wide open area, away from vents, cracks, and crevices in walls.
7. If there is a greater risk of fire, have a friend stand by with a fire extinguisher.
8. It is a good practice to stay in your shop one-half hour after all cutting and welding is done to make certain a fire is not smoldering somewhere.
9. Torches, plasma cutters, and carbon arcs do a great job of cutting though structural members. Think about the consequences of your actions before cutting. Take a minute to ask yourself what will fall down after the cut is made, especially when large or heavy sections of metal are involved.

SEALED CONTAINERS

One of the most dangerous situations is cutting into a drum, barrel, or tank without knowing what was or is inside. Avoid cutting, welding, or any hot work on any container that may have toxic, corrosive, reactive substances or flammable and explosive vapors. Don't rely on your nose to detect the presence of these substances. A small amount of residual flammable liquid or gas can cause a serious explosion. Reactive chemicals that have no odor at room temperature may give off large amounts of toxic vapors when heated.

Before cutting or welding a sealed container, the container must be cleaned. There are several methods for cleaning containers, depending on what was in them. For example, cylinders used for refrigerant can be made safe by having an HVAC professional apply a vacuum to it, sucking out the residue. Otherwise the refrigerant will turn into a highly poisonous gas when heated. I made the mistake of cutting into an uncleaned refrigerant drum. After plasma cutting just a few inches, a green-colored cloud of chlorine-smelling toxic gas came billowing out. All I could do is stand back and wait for the chemical reaction to stop. No one was hurt, but it was embarrassing having to explain to others why they needed to stay out of the area. An exact identification of the substance

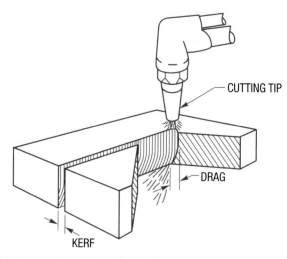

This diagram shows the oxy/fuel cutting process. Kerf refers to the amount of metal removed during the cut. *AWS A3.0:2001, Figure 41, Reproduced with permission from the American Welding Society (AWS), Miami, FL USA*

is required before cleaning, and only qualified persons should designate the cleaning procedure to assure it is carried out safely. Unless you are willing to do the required research and take all the proper precautions, it may be best to avoid this kind of cutting in the first place.

TORCH CUTTING (OXY/FUEL CUTTING—OFC)
How it works

When oxygen comes in contact with metal, a chemical reaction occurs called oxidation. Rust is a common form of oxidation and develops over a period of time. Torch cutting is a rapid form of oxidation and happens almost instantly. Torch cutting is a chemical cutting process, requiring no friction or mechanical force. The oxygen used for cutting

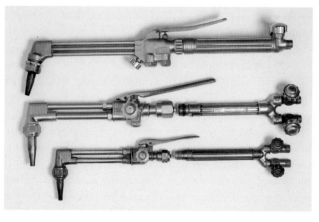

The Smith cutting attachment at the top has a 60-degree angled cutting head. The two Victor torches below attach to different-sized torch bodies. Other torches can be purchased that connect directly to the gas hoses. *Monte Swann*

must have a high purity. There is only 21 percent oxygen in the atmosphere, so torches that combine a fuel gas with air will not generate the amount of heat required for cutting. Oxygen cylinders contain gas that is 99.5 percent or 99.9 percent pure. If the oxygen purity is below 95 percent, cutting action disappears, resulting in melting and washing away the metal instead of cutting it. Read the chapter on oxy/acetylene and compressed gases for information on the different types of fuel gases, and equipment used in torch cutting.

Dross is the correct name for the byproduct of torch cutting, which is the oxidized steel. Dross is more commonly (and incorrectly) referred to as slag. Note that slag is actually the byproduct of flux used in welding or brazing.

Safety
Oxy/acetylene safety is covered in the same chapter that describes the different fuel gases and equipment. The precautions and guidelines for equipment safety are the same in gas welding, brazing, heating, and torch cutting. The type of clothing and eye protection—a No. 5 shade lens—is also the same. Torch cutting near concrete, stones, or block can cause the moisture inside to turn to steam and explode, sending pieces flying in every direction. If you have to cut close to these materials, put sheet metal down to deflect the heat.

Applications
Whether you are using acetylene or propane for your fuel gas, torch cutting is limited to carbon steel. All other metals cannot be cut, but can be melted with a torch. Steel is unique in that the melting point of the oxidized metal is lower than that of the base metal, allowing the cutting action to take place. With all other metals, such as stainless steel, brass, and aluminum, the oxidized metal created has a higher melting point than the base metal, preventing the torch from making a cut. Fortunately, carbon steel is the most common material used in metal working, so torches come in handy on a regular basis.

Of all the cutting processes, torches put the most heat into the base metal. Because of this, thinner pieces of steel are not typically torch cut because they tend to warp and distort from the heat. A pretty good cut can be made on sheet metals with a very small tip, like a No. 000 Victor, but this will still put a large amount of heat to the base metal. If you need to sever thin pieces, tilt the torch 45 degrees in the direction of travel and make the cut fast. With thin pieces of steel, consider the variety of mechanical methods of cutting, such as having pieces sheared to size at the steel yard. It may be difficult to torch cut through stacked materials, especially in the middle of a piece without an edge to start on. Stacked materials can be severed, but a clean straight cut is not likely to happen.

Torch tips
The tip of your torch is the business end of the cutting operation. Tips used with propane gas are two-piece tips.

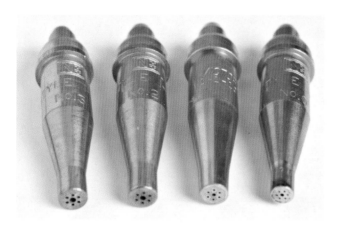

Invest in a variety of different-sized tips for your torch. The No. 3 tip, on the left, has the largest oxygen jet and preheat holes allowing it to cut through thicker metals. The No. 0 tip, on the right, is rated to cut through ⅜-inch maximum thickness. *Monte Swann*

Since propane burns at a lower temperature, the preheat flames are larger than those used with the one-piece acetylene cutting tip Be certain to keep your tip clean and in good condition. Before using any torch, even your own, clean all the preheat holes. This is especially important for the center hole, where the oxygen jet will blast out when the cutting lever is depressed. When using various sized reamers, start with the small sizes and work your way up. Forcing a tip cleaner into a hole may result in the reamer breaking off.

Tips come in a variety of sizes. Like welding, the tip size you select for cutting will depend on metal thickness. Larger-size tips can be used to cut thinner metals. But when metals are thicker, it becomes critical to use the proper-size tip for a successful cut. A tip chart is essential to selecting the right torch tip for the metal thickness and also gives the proper oxygen and acetylene pressure settings for each tip. Setting the pressures too low for the tip being used can starve the torch and lead to a flashback.

Scarfing tips are available for gouging applications with oxy/fuel. These tips are angled 90 degrees for ease in oxidizing and blowing metal off the surface. These tips tend to backfire and flashback when not used properly. A straight cutting tip can also be used to remove weld metal, trim up previously cut edges, and widen slots and holes. Remember, it is easier to leave extra metal when cutting and finish up with a grinder, than to overcut and have to make a major repair.

Flame adjustment

Lighting and adjusting the flame on a cutting torch is similar to a welding tip. Open the acetylene first and light the torch with a striker. It is possible to adjust the amount of acetylene up or down. When oxygen is added, it will vary the size of the preheat flames (cones). Smaller preheat flames can be used to make better cuts on materials thinner than the thickness range of the tip being used. If the cones are made too small,

preheat may take a long time, or the torch will pop out (backfire) because of the low flow of gas. Large cones will aid in cutting through metals on the high side of the thickness range. If the preheat cones are made too large, they will jump off the end of the torch tip, and remain as a carburizing flame. Preheat flames not only preheat the metal before the cut, but also preheat the metal during the cut. If the preheat holes are clogged, missing or deformed preheat flames can cause problems when making a cut. A neutral flame should be used for cutting. Adjust the final neutral flame with the cutting-jet lever depressed.

Cutting technique

Before you begin cutting, make certain that your equipment is set up properly.

Next, take the time to clean your metal. Paint, dirt, rust, and other contaminants will interfere with the cut. Paint can be burned with the torch and brushed off. Protect yourself from the toxic fumes if you decide to do this. When possible, clean both sides of the metal. Contaminants on the backside can stop the cut from going all the way through the metal. If only one side is slightly rusty, make the cut with the rusty side facing down and the clean side up.

Light the torch and adjust for a neutral flame. Preheat the metal at the beginning of the cut to a cherry red, or about 1,500 degrees F. When the metal is preheated, depress the oxygen cutting-jet lever all the way. Steadily move the torch in the direction of the cut, traveling as fast as you can without losing the cut. Keep the preheat cones about 1/8 inch away from the base metal. Check to make sure the sparks and dross are coming through on the backside. When the cut is finished, let off the cutting jet first, then move the torch away.

Torch cuts can be made in the horizontal, vertical, and overhead positions on sheet, plate, and structural shapes such as pipe. When making vertical cuts, try traveling downhill on thinner metals and uphill on thicker metal.

When adjusting the flame, depress the oxygen jet-cutting lever. If a small feather appears, increase your oxygen until a neutral flame is achieved with sharp, well-defined cones. *Monte Swann*

Grasp the torch lightly. Use one hand as a balance and pivoting point, and the other to depress the cutting-jet lever. For straight cuts, hold the torch square to the plate in all directions. If you tip the torch, the edge of your cut will have an angled face. *Monte Swann*

When using a cutting attachment, the oxygen valve on the torch body is fully opened. The flow of oxygen is controlled by the needle valve on the cutting attachment. *Monte Swann*

Hold the preheat cones directly over the edge where you plan to start cutting. Wait until the material turns a cherry red before proceeding. *Monte Swann*

Use a sharp piece of soapstone and combination square to make straight lines on the pieces to be cut. Soapstone is used to mark metal because it will not burn off. *Monte Swann*

Move off the edge of the material. Depress the cutting-jet lever and begin your cut. Because the edge of the metal was so hot, a small amount of dross has stuck to the beginning edge of the cut pictured. *Monte Swann*

When cutting pieces to size, the placement of your cutting tip is critical. Use the soapstone mark as a guide to keep one edge of the cut straight. If you cut down the center of the soapstone mark, your final piece may be undersized. *Monte Swann*

A circle burner is an adjustable jig for a torch used to cut precise circles of the same or different diameters. Magnetic brass straight edges and other jigs will aid in making smoother cuts than can be made by hand.

Troubleshooting

Plenty of things can go wrong, leading to a rough cut or no cut at all. If you are not making a good cut:

1. First, check the tip. A dirty tip is a big problem and is easily fixed. A dirty tip will deform the preheat cones, slowing down the cut. If the center hole is dirty, it will swirl the oxygen jet, gouging into the sides of the cut and slowing down progress.

I have moved my cutting tip to cut on the other side of the soapstone line. Changing the position of the kerf makes a difference in the size of the final part. *Monte Swann*

To cut a hole or shape in the center of a plate, preheat the metal like you would with any cut. More preheat is required since the heat is absorbed by the surrounding metal. *Monte Swann*

When you depress the oxygen cutting jet, dross will shoot up until your oxygen jet cuts all the way through the metal. To prevent your tip from becoming contaminated, move it back and forth slowly while depressing the cutting jet. Another method is to tilt the torch 20 degrees, blowing the dross to one side. To begin cutting, tilt your torch back to a 90-degree angle. *Monte Swann*

At this point, the oxygen jet has cut all the way through the metal. I can begin making the cut out. For large holes and slots, travel to the inside edge of the soapstone line and cut the shape. *Monte Swann*

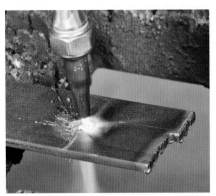

By making crosshairs on the metal, I can easily cut small holes, keeping them centered and round. *Monte Swann*

Save drilling time by using a torch to flame-cut bolt holes. Use scrap metal to practice this skill. *Monte Swann*

2. If you're not cutting all the way through the metal, or lose the cut at anytime, stop what you are doing. Dross coming up through the cut is always a bad sign. Check the amount of oxygen pressure. Too little pressure for the metal thickness will not allow the jet to penetrate all the way through. Too much oxygen pressure can cause the cut to widen on the backside.

3. Traveling too fast or suddenly switching the direction of the cut will not allow the preheat flames to do their job. Remember that they preheat the metal during the cut and need to be large enough for the metal thickness. Travel slowly enough to allow the preheat flames to work properly. Thick sections of metal can be slightly preheated the entire length of the cut to speed up the progression of the cut. Metals below room temperature should be preheated overall before the cut is made. Preheat flames are too large when the top edge of your cut melts into a radius, instead of being a sharp edge.

4. If the dross gets in the cut and is allowed to cool slightly, it will be difficult to cut back through it. You can make another cut beside the previous cut to avoid gouging into the base metal.

5. Some dross will be present on the backside of the cut. It should be easy to remove with a chipping hammer by hitting the dross back toward the edge of the cut. If the metal is allowed to get too hot, large amounts of dross will stick to the backside of the cut. Reduce your preheat time and increase your travel speed to compensate.

A handheld guide used to make a straight bevel cut. The guide is made of 1¼-inch outer-diameter round tubing with a small, square bar handle. Other straight edges can be made by welding Vise-Grip clamps to the end of scrap tubing. *Monte Swann*

A little more skill is required to use the torch in this way since one hand is used to hold the straight edge. Become comfortable cutting with both hands before using the torch this way. *Monte Swann*

The travel speed for cutting a bevel is slower than making a straight cut because a longer cross-section of metal is being cut. *Monte Swann*

Figure 14.22: Typical Edge Conditions Resulting from Oxyfuel Gas Cutting Operations: (1) good cut in 1 inch (25 mm) plate—the edge is square, and the drag lines are essentially vertical and not too pronounced; (2) preheat flames were too small for this cut, and the cutting speed was too slow, causing bad gouging at the bottom; (3) preheating flames were too long, with the result that the top surface melted over, the cut edge is irregular, and there is an excessive amount of adhering slag; (4) oxygen pressure was too low, with the result that the top edge melted over because of the slow cutting speed; (5) oxygen pressure was too high and the nozzle size too small, with the result that control of the cut was lost. *Welding Handbook Volume 2, Eighth Edition, Figure 14.22, Reproduced with permission from the American Welding Society (AWS), Miami, FL USA*

Figure 14.23: Typical Edge Conditions Resulting from Oxyfuel Gas Cutting Operations: (6) cutting speed was too slow, with the result that the irregularities of the drag lines are emphasized; (7) cutting speed was too fast, with the result that there is a pronounced break in the drag line, and the cut edge is irregular; (8) torch travel was unsteady, with the result that the cut edge is wavy and irregular; (9) cut was lost and not carefully restarted, causing bad gouges at the restarting point. *Welding Handbook Volume 2, Eighth Edition, Figure 14.23, Reproduced with permission from the American Welding Society (AWS), Miami, FL USA*

PLASMA ARC CUTTING (PAC)
How it Works

Instead of being a solid, liquid, or gas, plasma is a form of gas that has been destabilized by electrically heating it to a very high temperature. This ionizes the atoms and allows them to conduct electricity, which is referred to as ionized gas. This same plasma stream is also present in the GTAW welding process. Compressed air or bottled nitrogen is used, along with an electrical current to create plasma. Plasma actually forms in the small space between the electrode and tip of the torch. The orientation of these torch parts constricts the flow of gas, creating plasma, which greatly expands in volume and pressure. The small hole in the tip concentrates the heat and pressure of the plasma arc in a very small area. The plasma arc can reach 30,000 degrees F—hot enough to liquefy any metal—while the high pressure blows the molten metal away.

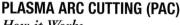

A Hypertherm plasma cutter with a 27-amp maximum output was used for these exercises. It can be used with either 110/115 or 220/230 input volts. This plasma cutter will cut through 18-gauge sheet metal at 160 IPM and 10 gauge at 57 IPM. The travel speeds drop off dramatically when you cut thicker metals—¼-inch metals are cut at 18 IPM and ⅜-inch-thick at 12 IPM. *Monte Swann*

Safety

The voltage used in plasma cutting is higher than most welding processes. In GMAW short-circuit transfer, the working voltage is between 15 and 22 volts. In SMAW, running 150 amps will mean an output of about 30 volts. Plasma-cutting machines run at around 50 volts. It takes less than one ampere to be killed by electrocution. Since most power sources run using more than one amp, the amount of voltage becomes the important risk factor. Higher voltages mean more force behind the electrical current and more chances of getting hurt. The plasma arc is hot enough to liquefy metal, and anything else it touches, including you.

Take some precautions when using this equipment. Make certain the ground clamp is securely fastened to the work piece, as close as possible to the cut. Turn off the machine whenever you move the ground clamp, or when changing out the parts of the torch. When you are finished cutting, turn off the machine and remove the ground. Plasma gas is not harmful by itself. Because plasma can cut through such a wide range of metals, consider the hazards associated with what you are cutting. Paint, oil, grease, and galvanized zinc coatings give off toxic fumes when burned. In any case, keep your head out of the smoke plume. Protect your eyes from the plasma arc light. Wear a No. 5 shade lens if you are not going to be looking directly at the arc. Wear a No. 8 shade when looking directly at the light.

Applications

Plasma cutting works on a wide variety of materials including steel, stainless steel, brass, aluminum, and titanium. Metals conduct electricity, and any electrically conductive material can be plasma cut. Plasma has a much lower heat input than torch cutting. Thinner materials, such as small-gauge sheet metals and thin wall tubing can be plasma cut with a minimum amount of distortion. Not only is the overall heat input lower due to faster travel speeds, but plasma also leaves a small heat affected zone. Stacked metal can also be plasma cut. All the pieces will have clean edges, as long as the power source is adequate to cut through the combined thicknesses. Gouging can also be performed with a plasma arc with the correct tip, electrode, and machine capacity.

Power sources

One of the limiting factors in using plasma for cutting is the high cost of purchasing a new machine. Smaller 12-amp machines are available and will sever cut ¼-inch material. This means the cut will be rough and travel speeds extremely slow. Larger machines allow the operator to make smooth, steady, good-quality cuts on 1-inch material, but cost as much as an industrial capacity GTAW inverter machine. To make an educated decision on purchasing a plasma-cutting machine, decide what types of metal and metal thicknesses you will routinely cut and how fast you want to make those cuts.

The speed at which you can cut will have a direct effect on the quality of the cut. Traveling at a speed of 10 IPM is the recognized minimum for a quality cut. However, even 12 to 18 IPM is a slow travel speed, taking more time to make the cut with a lot of heat reflecting back onto your gloves/hands. With a travel speed of 50 to 60 IPM, you can

The plasma cutter is shown cutting through brass, stainless steel, copper, and aluminum sheets stacked together. Making this cut required the full capacity of the machine. *Monte Swann*

Connect the air hose before turning on the power source. Notice all of the warning stickers on the top of the machine. Plasma cutters are high-voltage equipment that can be dangerous and should be used with caution. *Monte Swann*

make a reasonably quick cut at a comfortable speed. Judge the plasma cutter by the thickness of metal it will cut at 60 IPM instead of 10. Consider the total output of the machine (amps × volts = watts), as well as the cost of replacing the torch consumables. Tips and electrodes are replaced on a regular basis. Plan for the continued expense; find out how much these consumables cost before purchasing a machine.

An air compressor is needed for plasma cutting. The compressor must have enough capacity to maintain constant pressure (about 70 to 90 psi) throughout the cut. Some plasma-cutting power sources will have an air compressor built into the machine, making it convenient to operate and more portable. Maintain your compressor and filters so the supplied air is dirt and moisture free.

Plasma torches

The most common consumables on a plasma torch are the tip and electrode. The electrode is made of copper, with a tiny rod of a rare metal—hafnium—in the center. Like the name implies, consumables are used up frequently during cutting and are replaced often. Consider the cost of replacing the tip and electrode, which last for about 15 minutes of arc on-time. Certain circumstances lead to an even shorter life for consumables, such as dirty or moist compressed air. Cutting perforated or expanded metals, where the arc is constantly turning on and off, will also shorten their life span. Long, steady, controlled cuts will extend the life of consumables.

Cutting technique

Refer to your manual for suggested amperage and air pressure settings. Most new machines will have a trigger lock to prevent accidentally starting the plasma arc. If you are using less than 40 amps, like when cutting thinner metals, the plasma torch tip can be placed directly on the metal during the cut. That way you will not need to look directly at the arc and can use a No. 5 shade. When running more than 40 amps, the torch will need to be held about ⅛ inch away from the metal.

Maintaining this gap is the most difficult part of making a good cut. Drag shields or standoffs can be used at the end of a torch to maintain that gap. Some of these accessories will even have small wheels for more control during a beveled cut. Before making a cut, get the plasma torch positioned close to the work piece. With your shade down, depress the trigger and make the cut. Be certain to let off the trigger at the end of the cut before you pull away, otherwise you will gouge into the top of the material.

With the power off, I have removed the tip and electrode from the plasma-cutting torch. The consumables are worn, discolored, and need to be replaced. This torch is a newer model with a yellow, flip-up trigger guard to prevent accidentally starting the plasma arc. *Monte Swann*

Begin making your first cut by holding the torch on the edge of the metal and depressing the trigger. *Monte Swann*

When the hafnium electrode becomes pitted more than ¹⁄₁₆ inch, it needs to be replaced with a new one. *Monte Swann*

Intricate cuts can be made with a plasma torch on very thin-gauge materials. *Monte Swann*

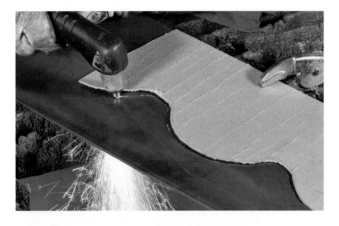

To pierce a hole with a plasma torch, tilt the torch 20 degrees, depress the trigger, and wait for the plasma arc to cut through the metal. *Monte Swann*

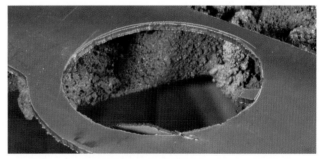

Once the torch has cut through and the sparks are coming out the back, travel to the inside edge and begin making your shape. *Monte Swann*

Cardboard or plywood shapes can be used as templates for cutting metal shapes. The edge of the template is used as a guide for the plasma-cutting torch. The template should be slightly smaller than the final part to compensate for the width of the torch tip. *Monte Swann*

This piece is shown after the cut has been completed. Unlike oxy/acetylene torches, plasma cutters can leave smooth, high-quality cut edges on thin pieces of metal. *Monte Swann*

Troubleshooting

The biggest issue with plasma is not cutting all the way though the material. If this happens, make adjustments to the amperage and air pressure setting. Increase the amperage only when there is enough air pressure being supplied for the cut. Next, check the tip and electrode. Double-check your ground connection. Make certain there is good contact between the ground clamp and work piece. Worn consumables will widen the cut and reduce the working pressure. Finally, reduce your travel speed so that the molten metal is blowing out of the backside of the cut. Any erratic or sudden movements will cause the cutting action to stop, especially when cutting metal thicknesses at the high end of a machine's capacity. Let off the trigger, reposition yourself, and begin again. Unlike oxy/acetylene, plasma arcs will cut back through areas where dross is present.

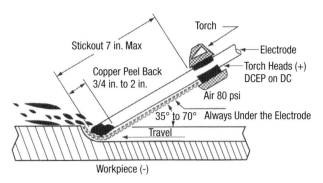

Welding Handbook Volume 2, Eighth Edition, Figure 15.4, Reproduced with permission from the American Welding Society (AWS), Miami, FL USA

CARBON-ARC CUTTING (CAC)

Carbon-arc cutting works by sending electrical current through an electrode made of solid carbon. The heat of the arc at the end of the electrode liquefies the metal and a high-pressure stream of compressed air blows large quantities of dross away.

Safety

Carbon-arc cutting is extremely loud. Hearing protection is a must with this process. Review all the hazards in chapter 9. Unlike torch cutting and plasma cutting, the carbon-arc electrode removes far more metal, creating a high volume of sparks and a large amount of hot dross. A welding helmet with a No. 10 shade lens will need to be worn. Darker lenses may need to be used for higher amperages.

Applications

The carbon-arc process is more likely to be used for gouging operations than cutting. Carbon-arc gouging is used to remove unwanted weld beads or when back-gouging a joint in preparation for complete joint penetration welds. The carbon-arc electrode also leaves a unique radius groove when it is used for edge preparation on plate stock. J-groove and U-groove joints are prepared by CAC operations. A variety of materials can be cut and gouged with the carbon-arc process including carbon steels, stainless steel, aluminum, cast iron, copper-based alloys, nickel alloys, and magnesium. Since the electrode is made from solid carbon, a small amount of carbon will be transferred to the base metal.

After gouging metals such as aluminum, brass, and even steel, the gouge will need to be sanded out to remove any carbon residue before welding. Since a CC SMAW/GTAW power source is used, carbon-arc cutting and gouging can be another way to utilize this equipment. A consistent supply of compressed air is required along with a carbon-arc electrode holder. The SMAW electrode holder (stinger) and compressed air hose attach to the electric lead of the carbon-arc torch. A standard ground clamp is used to complete the circuit.

Electrodes

The carbon electrodes come in a variety of diameters. The larger the diameter, the more amperage output is required. Eighth-inch diameter electrodes operate between 60 and 90 amps on DCEP, while a ¼-inch-diameter electrode requires about 300 to 400 amps. Use electrodes that match the output capacity of your welding machine and equipment. The diameter of the electrode will also determine the depth and radius of a gouge it can make. Deeper gouges can be made with smaller electrodes by making multiple passes. The copper foil around the outside helps conduct electricity to the cutting arc.

Technique

Once the machine controls have been set to the proper polarity and amperage, secure the ground clamp to the work piece. Place the electrode in the holder, sticking out 3 to 4 inches, or the width of the palm of your gloved hand. Be certain the small holes for the compressed air jet are pointing in the direction of travel and are between the base metal and the electrode. Turn on the air, get into position, and lower your helmet. You do not need to strike an arc as in stick welding. Start the arc simply by making contact with the metal.

Do not pull the electrode away from the work. After the arc has started, maintain a 35- to 45-degree travel angle and push the electrode along as fast as it will cut into and blow away the base metal. This will be a faster travel speed than in SMAW; traveling too slow will extinguish the arc. The electrode will be slowly consumed in the cutting process; don't allow the electrode to get too far away from the cut.

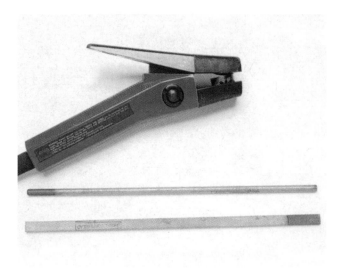

A carbon arc electrode holder is at the top. The button on the handle is depressed to activate the compressed air stream. The two electrodes at the bottom are round and rectangular for different cutting applications. *Welding Inspection Technology, Fourth Edition, Figure 3.64, Reproduced with permission from the American Welding Society (AWS), Miami, FL USA*

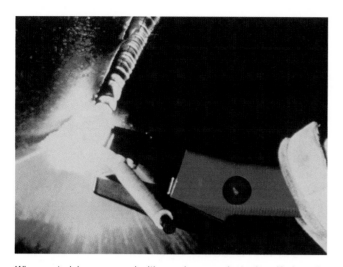

When materials are gouged with a carbon-arc electrode, a U-shaped groove is created in the metal. *Welding Inspection Technology, Fourth Edition, Figure 3.63, Reproduced with permission from the American Welding Society (AWS), Miami, FL USA*

The depth and contour of the groove is determined by the electrode diameter and travel speed. The general rule is groove depths should be no deeper than 1½ times the electrode diameter. Deeper grooves and cuts are made with multiple passes. Groove width will be about ⅛ inch wider than the electrode; a wider groove can be made by weaving the electrode or using a small circular motion. The procedures for cutting are the same as gouging, only the electrode is held at a much steeper angle, 70 to 80 degrees from the plate. The only time the electrode is held straight up and down is for piercing holes.

Automated cutting

Most shops, both large and small, will have some type of automated cutting system. If you need a large, intricate part cut precisely, or multiple parts that are all the same size, check out the job shops in your area. A job shop will take orders from the general public and, for a fee, fabricate the part to your specifications. You will need to supply a drawing with the location and sizes of all features. If you are familiar with CAD (computer automated design) or know someone who is, a DXF file of your drawing is the best way to give them your information. Otherwise, there may be additional programming costs if the drawing is given in another format, like a simple sketch.

When specifying tolerance, remember what is important for the fit up of your project and what is not. The more precise the part, the more it will cost. Drilled and tapped holes may need to be located within 0.001 inch so the mating parts will fit up properly; this is a common tolerance in machining operations. For parts that are to be welded a tolerance of ±0.005 inch is very precise. With mechanize cutting, a tolerance of ±¹⁄₁₆ inch is very reasonable. For manual

OFC, a tolerance of ±⅛ is expected. For mechanized cutting, tolerance will depend upon the method. Flame cuts made with an automated oxy/acetylene torch will be less precise than parts cut with a CO_2 laser. Plasma torches and water jets are also used in automated cutting operations where parts are cut with the assistance of motorized tracks and CNC.

Heat input is another factor to consider. Water jets utilize abrasives in a high-pressure stream, more akin to a mechanical cutting method. Mechanical cutting methods, including water jet, leave no heat-affected zone in the finished material. Even materials such as rubber and glass can be cut precisely with a water-jet system. The heat input from laser cutting is very minimal. Plasma cutting puts a little more heat into the materials, but still leaves a very small HAZ. Torch cutting, by far, puts the greatest amount of heat into a material and is usually only used in automated cutting on thick sections of steel (more than 1 inch).

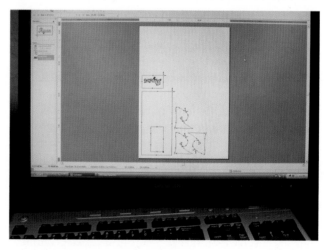

Automated cutting is commonly used in manufacturing to cut parts. This CNC plasma cutter is running through a program downloaded from a personal computer. The computer monitor shows how nesting software has automatically calculated the most efficient way to cut out all the parts. *Monte Swann*

CHAPTER 14
FABRICATION AND ADDITIONAL EXERCISES

When building a project, it is typical to spend 90 percent of the time cutting materials to size, fitting them together, cleaning the joints to be welded, and tacking up the pieces. The last 10 percent is spent welding the joints. Think of these as two different skill sets: fabricating and welding. A majority of the time can be spent fabricating a beautiful project only to mess it up with bad welds. A poorly fabricated project with misaligned joints, large gaps (root openings), and poor fit-up will be difficult to tackle for even the most skilled welder.

This book has discussed the basics of safety, use of equipment, joint designs, and welding technique. Start with the basics. Learn how to weld before diving into a major project. Observe the behavior of the weld metal and base metal while practicing your technique. Once you see the effects of heat and welding, you will become a much better fabricator. Welding a project together is the most critical. It is the easiest to get wrong and the most difficult to fix. Understanding residual stresses in the base metal and the shrinkage stresses from welding are key factors in being successful with 100 percent of your project. Although we briefly touched on some of the following topics, it is important to review them as they relate to distortion.

CAUSES OF DISTORTION

There is no getting around the fact that heat is required for manual welding processes. When a chunk of metal is heated uniformly, it expands in all directions. When it contracts, it shrinks back very close to its original size. If the entire piece of metal is heated, the stresses will become equalized and no distortion will occur. When a bar is heated on one side, the hotter side to which the torch is applied will expand. Because the heated side expands more than the unheated side, stresses occur between the very hot metal and surrounding metal.

As the metal expands and contracts, it puts some areas of the metal in tension (heated part expanding) and other areas in compression (unheated part resisting the expansion). As the metal cools, these opposing forces gradually reduce down to zero. But this zero point is reached before the metal has completely cooled. Shrinkage occurs as the metal continues cooling to room temperature. Shrinkage stresses continue to cause tension in the weld, continuing to pull the surrounding metal towards the heated area. When this happens, it changes the shape of the piece causing distortion. With low-carbon steel, 200 degrees F is enough heat in an unrestricted part to cause distortion. At that temperature, the metal has been stressed beyond its yield point resulting in a change to its dimensions (plastic flow).

RESIDUAL STRESSES

All structural shapes have residual stresses, unless they are completely annealed. It is typical for residual stresses to be present. As metal is heated and rolled into bars, angles, channel, and sheets, the forces of tension and compression act upon the materials. As the metal becomes a finished structural shape, these residual stresses remain behind, trapped inside the metal. If the geometry of the shape is changed, the balance of residual stresses is changed, distorting a once straight piece of metal. For example, if an I-beam is sawed down the middle of the web, the two T-shaped sections will bend outward. Also, small strips sheared from plate or sheet will be twisted.

Accounting for the dimensional changes from residual stresses is important in machining operations, as well. A rectangular bar with a slot machined down the length of one side will cause the bar to curve or bow away from the slot. When the bar was made, the residual stresses were evenly distributed, keeping it straight. As material was removed from one side, less stress was concentrated in the remaining metal. The other side has the same amount of stress as before. The uncut side now has more residual stresses than the cut side. The material is pulled toward the side with the most remaining residual stresses.

FIT-UP

Two pieces of metal can be put together in ways that make welding easy or difficult. Joints with poor fit-up are difficult to weld. Large gaps require large amounts of weld metal to fill them. The more filler metal added to a joint, the more shrinkage stresses occur. The more shrinkage stresses, the more distortion of the final piece. Do a good job fitting the pieces together, keeping the edges straight and square. If necessary, build up metal on one piece before welding it to another.

NUMBER OF PASSES

Thicker metals require multiple passes to fill a weld joint or make a large-size fillet weld. Make the fewest number of passes possible by using larger-diameter electrodes. The shrinkage caused by each pass tends to be cumulative. The more passes made in a joint, the more distortion will occur.

WELD PROFILES

Fillet welds with excessive convexity, or groove welds with excessive reinforcement, create a notch effect in the part. The extra weld metal does not make the weld any stronger. Having reinforcement on a groove weld is standard practice, but the height of reinforcement should not exceed ⅛ inch. Unnecessary filler metal will also increase the amount of shrinkage stresses and distortion. Weld beads with a flush, or slightly convex contour, have a lower amount of distortion. Concave profiles will have the least amount of distortion on the surrounding base metal. However, with concave profiles, the solidifying weld metal is in tension and will tend to crack in the crater of the bead.

WELD SIZE

Fillet and groove welds do not need to be larger than the thickness of the base metal. Large welds will not make the joint stronger. With large welds, the shrinkage stresses are increased by the large amount of filler metal, causing excessive distortion. Wide bevels (included angle over 60 degrees) require more weld metal to fill the joint, putting a larger volume of heat into the base metal.

NEUTRAL AXIS

Defined as the center of gravity, the neutral axis is the balance point of a weld joint and part. Welding near the neutral axis reduces the amount of distortion, since each side has an equal mass of metal to absorb the heat from welding. For example, take a large metal spoon and balance it on one finger. This would be the best place to cut and weld the spoon back together with the least amount of distortion. Welding the spoon from one side with no gap will still result in higher shrinkage stress pulling the base metal in toward the weld. Putting a root opening in the joint will distribute the weld metal more evenly through the center (thickness). Root openings reduce the amount of shrinkage stresses pulling the joint to one side. Welding the spoon on both sides reduces the distortion in a similar way. Think of this example when designing joints for your project.

DISTORTION CONTROL TECHNIQUES

1. *Don't overweld.* Most beginners make the mistake of using too much weld metal on a project. The welds made in this book are the length of the joint *only because they are for practice.* A ½-inch-long bead at either end of the 6-inch-long joint would be more than enough to hold the pieces together. There are situations where a full-length weld is required—for example, any part that needs to be air/watertight such as a fuel tank or small diameter pieces of tubing. In most applications, full-length welds are not necessary. Many electrodes and filler rods have a tensile strength of 60,000–70,000 psi. One square inch of filler metal will hold 30 tons if welded properly. A 2 × 2-inch bar would hold 240,000 psi tensile. Only you will know the service requirements for the part. As a general

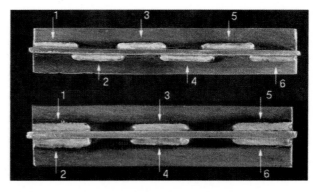

This photo shows examples of intermittent welds. Above is a staggered intermittent weld. Below is a chain intermittent weld.
Welding Inspection Technology, Fourth Edition, Figure 10.11, Reproduced with permission from the American Welding Society (AWS), Miami, FL USA

thought, be conservative with the amount of welding on a project. If the joints are welded properly, a part may need less filler metal than you think. The larger the amount of filler metal, the more stress in the solidifying weld metal and distortion in the pieces.

2. Back stepping and intermittent welding can spread out the heat. Distortion is minimized because there is never a localized buildup of heat resulting in increased shrinkage stresses.

3. Pre-bending joints is a way of compensating for distortion. Joints are preset at certain angles, so when welding is complete, the shrinkage stresses in the weld metal have pulled the pieces into alignment. This method of distortion control takes experience to be used effectively. Try tacking a tee joint together, about 5 degrees off of the usual 90. When the parts are unrestricted by clamps, the shrinkage on fillet welds pulls the base metal about 1/32 inch per weld pass. There will be even more pull if the fillet-weld leg size exceeds ¾ of the base-metal thickness.

4. The direction of welding should move from the most restrained area toward a less restrained area. The principle applies to all the joints being welded. For example, think of a simple frame with a few cross pieces in the center. The center pieces are more restrained by the structure of the frame. If the outside perimeter of the frame is completely welded first, the pieces in the center will have more residual stress after welding. If the center pieces are welded before the perimeter of the frame, they are less restricted by the surrounding metal. Because they are allowed to move somewhat during welding, less residual stresses will be left in the welded joints. On a complex piece, start with the most restrained pieces. They are usually located in the center, surrounded on several sides by other parts. Move out from the center toward the edges where parts have a greater relative freedom of movement.

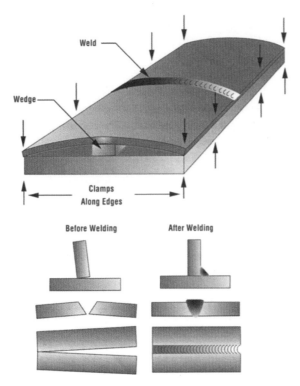

Weld

Wedge

Clamps
Along Edges

Before Welding   After Welding

Welding Inspection Technology, Fourth Edition, Figure 10.6,
Reproduced with permission from the American Welding Society
(AWS), Miami, FL USA

5. Use mechanical restraints, jigs, fixtures, and/or clamps. These tools can be used for pre-bending a joint or holding it in place while it is being welded. Clamping parts to a heavy plate (table top) or large structural shape, such as an I-beam or piece of channel, will provide the needed support. Restraining the pieces through a heating and cooling cycle will reduce the amount of distortion. Keep parts secured through the important point where the shrinkage stresses have built up enough so that the yield strength of the weld is exceeded. When the yield strength is exceeded, the filler metal can bend and stretch.

If clamps, jigs, and fixtures are used, the restrained pieces will have a greater amount of residual stress locked up in the metal. Hot, restrained metal will still expand and contract, but is prevented from moving. As a result, there are higher levels of strain and more residual stresses when the metal cools. If the residual stresses build up to the point where they exceed the strength of the base metal, it will crack. This is one reason why some metals need to be stress relieved after welding. Post-weld heating is one of the most common methods of stress relief. This type of stress relief is typically done in an oven capable of uniform temperature regulation. Portable machines with heating coils are used to heat-treat parts welded in place.

If an unrestrained piece of metal is allowed to move during the cooling cycle, some of the internal stresses are relieved, but the piece is distorted. There is a compromise to the situation. Use just enough clamps and fixtures to hold the pieces in place, but not so many that the pieces cannot move at all. Preheating the entire part before welding will reduce the temperature difference between the base metal and weld area. As long as the amount of preheat is uniform throughout the piece, the rate of expansion and contraction is reduced. After welding, the part must be allowed to cool slowly.

DESIGN

Engineers spend many years in school learning about load, stress, and strain. The mathematics and theory is important when designing buildings, bridges, cars, and airplanes. Another type of knowledge is equally as important: learning through experience. The more projects you build, the more experience will teach you what works well and what doesn't. Don't expect every project to be a success. When something goes wrong, the important thing is that you learn something new. There are a few general considerations when designing a project:

1. Direction of load is applied to a welded joint. A joint welded on one side can withstand more force when it is applied to the face side. On a tee joint, the side without the fillet weld is the weakest. If you place the tee joint with the weld face-up and hit the top edge with a hammer, it is easier to fracture. With the weld facedown, the load applied to the top edge (via the hammer) will have to be dramatically increased to break the joint. If the weld is properly fused into the joint, the base metal is likely to bend before the weld breaks. Joints can withstand the most amount of force when loaded in the longitudinal direction. A tee joint in the vertical position will support a lot of weight (load) located directly above (weight on top being supported) or below the joint (weight hanging underneath).

2. Direction of load applied to a structural shape. A piece of round stock or tubing is a cylinder. A cylinder can withstand the greatest amount of load when it is applied in compression or tension to the round ends. For example, it is easier to break a tree branch by bending it, than by pulling it apart. The same principle is also true with metal shapes.

3. Orientation of structural shapes. Rectangular tubes can withstand higher loads when placed with the longer side in line with an applied force. For example, a piece of rectangular tubing is a better structural shape to use for a lifting bar (for a hoist) than a square piece of tubing.

4. Geometric shapes, like triangles, will support more weight than rectangles or squares. Look at a bridge or building truss as an example of how this theory is applied. Gussets are pieces of metal cut into a right triangle shape. Gussets are welded into corners (pieces

that come together at 90 degrees) to reinforce part of the structure.

5. Thicker pieces of metal can withstand more load than thinner pieces, no matter if the metal is a structural shape, such as tubing, or plate and sheet.

6. Select a base metal that is easy to weld. Exotic alloys need specific filler metals and welding procedures. Use a base metal that best suits your applications, but keep in mind certain metals are more difficult to weld than others. Whenever possible, select a steel within the normal range, such as AISI-SAE 1015 and 1025, or steels that contain under 0.35 percent sulfur, a maximum of 0.1 percent silicon, and 0.15 percent to 0.25 percent carbon content.

PROJECTS

Keep safety in mind at all times. Inspect the work area for any hazards and wear the PPE. General fabrication techniques can be used by hobbyists and professionals alike:

For this frame, notch a piece of angle iron and form 90-degree corners. Note that the angle iron was notched a different way in each of the corners. *Monte Swann*

Practice notching structural shapes, fitting pieces together, and welding them. Here, saw and torch cuts were used to fabricate with pieces of angle iron, square tube, and small-diameter pipe. The pieces were fitted together and welded with a 6011 electrode. This project was created for practice only, but could easily be something useful like a trailer hitch. *Monte Swann*

1. Have a plan. Take some time to make a few simple drawings. Think about the size of the project and how parts are going fit together. Experienced people do a lot of fabricating in their head. They can picture in their minds how the project will turn out when it is put together in different ways.

2. Decide on the structural shapes, sheet, or plate you want to use. Make a list of materials you need before going to the steel yard. Buy enough material so you have extra. That way, changes can be made during the project if your initial design does not work out.

3. Take time to cut the pieces accurately. Measure twice; cut once. Cutting a piece too large is easier to fix than cutting it too small. Joints should have good fit-up for ease of welding and less distortion.

4. Groove welds on thicker pieces of metal should be beveled for adequate penetration into the joint. This is especially important when the weld bead is going to be ground flush with the base metal.

5. Clean your metal. Joints and the area around them should be free of paint, rust, dirt, oil, and mill scale. A clean weld is a strong weld.

6. Select the best welding process for the job. If the machine or process you are using is underpowered for the job, avoid doing the work.

7. Match the filler metal to the base metal. If you are not sure which one to use, do some research to find the answer.

8. Tack all the parts together first. It is much easier to cut a tack than to cut a weld. Tacks are easily removed with a cut-off wheel, leaving the rest of the joint intact. The size of a tack should be in proportion to the metal thickness and size of the part. A tack on sheet metal can be small, while plates 1 inch thick may require large tacks 2 inches long.

9. Three tacks in a triangular orientation will hold pieces in place. Joints with one or two tacks can be pulled out of alignment when welded.

10. Measure the project for dimensional accuracy. Now is the time to make adjustments before welding. Since tacks are small, parts are easy to move into alignment with clamps. If a tack breaks, or you cut it apart, sand it down before retacking.

11. When using a welding table, use clamps to secure the pieces to the flat surface. Having a large number of Vise-Grips and other clamps comes in handy.

12. Make sub-assemblies. For example, if the project is a table, it may be useful to tack together each side first. Then, bring the two halves together with the cross pieces in between. On other projects, separate pieces can be completely welded before being joined to the whole project.

13. Get some scrap pieces of metal the same thickness as your project. Take the time to dial in the machine settings. Try to use the same joint designs you will be

welding. Setting the heat on a tee joint may end up being too much for an open-root groove weld.

14. Begin welding. Don't weld in just one area, skip around. Use back-stepping and intermittent welds to reduce the amount of distortion.

15. Wrap your welds around corners (boxing). Corners are points where stresses are concentrated. Reduce the chances of the weld cracking; do not end a weld bead on a corner. If a joint should be welded all the way around, don't leave any gaps or lack of fusion where the beads connect with each other.

16. Use distortion to your advantage. Keep in mind that the metal adjacent to a weld will pull toward the weld. Use this to your advantage when welding a frame that is not completely square. By welding the inside corners of the shorter side, it may pull the longer side into alignment. Metals will also be pulled in the direction of welding. If the root opening of a groove weld is wider at one end, start the weld on the other end. As you weld, the pieces may be pulled into alignment.

17. Stop welding if things are not working out. Troubleshoot any problems. You may be using incorrect equipment, machine settings, shielding gas, or electrodes. The problem could also be caused by improper welding technique. If the problem is lack of experience, you are in luck. A little more practice is all it usually takes to fix that problem.

WELDING SHEET METAL

Thin stainless-steel (300 series) sheet metals have to be welded with great care to minimize distortion. Since stainless steel expands more than other metals when heated, it will distort to a greater degree. Other metals, which cannot be hardened by heat treatment such as mild steel, can be welded with the technique outlined below. Since stainless steel distorts 50 percent more than mild steel, it is very challenging to make the finished piece come out right.

1. Make many small tacks, every 1–2 inches the entire length of all the joints.
2. Fit-up is very critical; the joints should be tacked tightly together.
3. Make 1-inch-long weld beads.
4. Cool the short weld beads immediately with a wet rag.
5. Move to another area away from your first bead. Make another 1-inch weld and cool it.
6. Skip to another area and continue this process. It takes time and patience to complete a project this way, but the distortion will be minimal if done correctly.

Using this technique increases the number of heating and cooling cycles, but reduces the amount of heat concentrated in one area. This reduces the amount of compression and tension stresses in the total volume of the base metal.

STRAIGHTENING DISTORTED PIECES

If your part is somewhat distorted from welding, there are a few methods to correct the problem:

1. Mechanical means. A hydraulic jack, clamps, pry bars, and levers can be used to move parts back into alignment. You may need to overbend the pieces slightly because they tend to spring back a little when the pressure or elastic strain is released.
2. Localized heating. Heating small areas with a torch can be done intentionally to bring parts into realignment. Heat can also be applied by running weld beads on the convex surface to pull the piece back into alignment. The beads can be ground off later for a smooth finish.
3. Localized heating used in combination with mechanical means.

Localized heating can be applied in a spot, line, V-shape, or block pattern on sheet and structural shapes. Keep in mind, thin-gauge metals are affected by the heat and are more likely to warp. If done correctly, non-heat-sensitive metals can be brought into alignment. Do not use heat-shaping methods on materials such as alloy steels; it could have detrimental effects on the mechanical properties.

1. Mark the area to be heated with soapstone.
2. Heat the spot to a specific temperature and begin moving in a pattern or straight line. Note: The metal will pull toward the side that is heated.
3. Keep temperatures below a red heat. Localized heating temperatures should not exceed 1,200 degrees F. Use temp sticks to precisely measure the temperature.
4. Do not backtrack over areas already heated.
5. Spraying the heated area with a fine mist of water may increase the shrinkage stresses and further distort the base metal. With localized heating, the distortion is used to pull the pieces into alignment.

ADDITIONAL EXERCISES

Round stock sign

An easy way to get started with metal fabrication is to make a sign with round steel rods. Rods up to ³⁄₁₆-inch diameter are inexpensive to purchase. A 12-foot piece is only a few dollars. Two or three different sizes of round stock can be used for added interest. The smaller diameter is easy to cut with a pair of wire or bolt cutters and easy to bend. The rods can be cold formed by hand bending or hammering, or hot-formed with the aid of a torch. Once the mild steel turns a red-orange color, the plasticized metal is very easy to bend and shape. Use two pair of pliers to handle the hot rods. Pieces of pipe and other meal shapes can be used as jigs to bend parts into round shapes or rings.

First, make a drawing of what your sign will look like. How tall will the letters be? How much space between them?

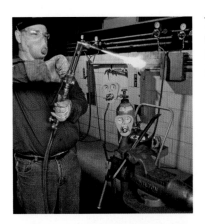

This crowbar is more useful to me if one end is bent 90-degrees. *Monte Swann*

This round rod sign was made from 1/16-inch-diameter, copper-coated, mild-steel filler rods. *Monte Swann*

Since the grains will stretch on the outside of the bend, heat the inside of the crowbar so the metal can easily stretch. Don't go overboard with the heat. Be conservative. Tools are usually hardenable by heat-treatment. A piece of tubing makes a great lever. If the crowbar does not bend, apply a little more heat with the torch. *Monte Swann*

That looks about right. Next, wrap the crowbar in a fire blanket so it cools slowly. If the crowbar is cooled in the air, or quenched in water, the steel may harden and become brittle. *Monte Swann*

Will there be a border around the outside? The drawing can be smaller than the project or made to scale. A drawing made to scale is the same size as the finished project. The life-sized drawing can be used as a template. As you are cutting, bending, and forming the rods, the dimensions can be checked with a full-scale drawing for accuracy.

Once all the pieces are fabricated, have a plan to join them together. With mild steel, any welding or brazing process will work. Remember, torch tips and electrodes are designed to work for a range of material thicknesses. The type of torch or welding equipment you have may be a factor in what size round stock can be used. It is possible to weld together very small-diameter rod stock, such as filler rods, with some finesse.

Next, make some practice welds on scrap materials. Make the same types of joint you plan on using in the sign. After all the parts have been joined, the project can be left as is. Or, sandpaper and a wire brush can be used to polish the metal. If the sign is for outdoor use, coat it with a rust-resistant paint or clear enamel spray to help prevent rust. If the project is made of stainless steel, aluminum, or if you want a weathered look, no rust-prevention treatment is necessary.

REPAIRING A HOLE

One way to improve your welding skills is to practice filling holes. GMAW short-circuit transfer, GTAW, and gas welding are good processes to use for this exercise because they don't use flux. SMAW can be used, but the flux will need to be cleaned from the weld between each layer to prevent slag inclusions.

Start by building up filler metal around the circumference of the hole. Keep turning the torch or gun so the heat of the flame or arc is directed back at the base metal. The thickness of metal will determine how much welding can be done before things get too hot. Thicker pieces can be welded all at once. Thinner sections will have to be built up gradually and allowed to cool periodically. With time and patience, any size hole on any thickness of material can be repaired by welding.

If there is a hole that needs to be filled on a welding project, a copper backing bar can be used. Steel, stainless steel, and aluminum base metals use copper or stainless-steel (for aluminum) backing bars. A copper backing bar clamped underneath makes it much easier to fill in the hole with weld metal. The filler metal will not stick to the copper, so the backing bar is easily removed after welding is complete. Copper backing bars clamped on the back side of a joint will draw heat away from the area being welded. Using a backing bar helps minimize distortion and can prevent melt-through.

Form a molten pool on one side of the hole. *Monte Swann*

Rotate the torch and add filler metal to the puddle. *Monte Swann*

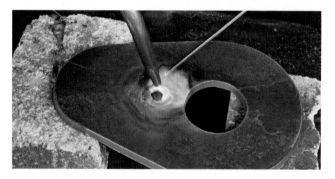

Keep traveling around the circumference until the hole is filled. *Monte Swann*

Now the part can be flipped over to fill the backside, if necessary. The large hole can also be filled in, although it may be faster and easier to cut a patch and weld it in place. *Monte Swann*

Since no flux is used with GMAW, holes are quickly and easily filled without the risk of slag inclusions. *Monte Swann*

Keep the wire directed at the base metal. Pull the trigger intermittently, adding a small amount of filler metal at a time. Keep moving and adding filler until the hole closes up. *Monte Swann*

Grind the reinforcement off for a smooth finish. The discoloration of the base metal around the weld is from the zinc coating on the material. *Monte Swann*

WELDING SHEET METAL WITH E6013

Because of its lower penetration characteristics, the 6013 electrode is designed to weld sheet metals at fast travel speeds. Use a 3/32-inch-diameter electrode with DCEN to weld 14-gauge, mild-steel sheet metal. Start with an amperage setting 40 percent less than what is recommended for the electrode. Refer to the SMAW chapter or electrode manufacturer for this information. Although the electrode is a larger diameter than the base metal (3/32 inch = 0.093 inch and 14 gauge = 0.074 inch), the low amperage and faster travel speeds reduce the amount of heat input. Smaller-diameter rods, such as 1/16 inch, can also be used, but the electrode will deposit a shorter length of weld bead.

Strike the arc on the end of the corner joint. Travel as fast as the filler metal in the electrode can be deposited in the joint. Move in a straight line; do not use a circular or zigzag motion. *Monte Swann*

Try not to hesitate. Keep moving until the end of the joint and pull out of the weld pool. Don't spend time filling in the crater. *Monte Swann*

The finished weld has complete fusion without excessive buildup. *Monte Swann*

THE CUBE

After you have become comfortable with a welding process and are familiar with how it works, try building this simple project. Making a cube is a great way to see firsthand how fit-up and welding techniques affect each other. Each of the six pieces for the cube needs to be as square as possible. Measure and cut or shear the pieces as accurately as possible. Clamp them together in a stack and take the time to file or sand down mismatched edges. Remember, joints with poor fit-up will be more difficult to weld.

Any welding process can be used to make the cube. With gas welding and GTAW, filler metal is optional. With GMAW, try making the welds in the vertical position traveling downhill for a smooth bead. Try making cubes of different sizes, with different gauges of sheet metal, on steel, stainless steel, and aluminum.

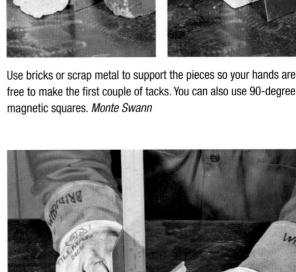

Use bricks or scrap metal to support the pieces so your hands are free to make the first couple of tacks. You can also use 90-degree magnetic squares. *Monte Swann*

Straighten the first two pieces. Use a combination square to check the angle. *Monte Swann*

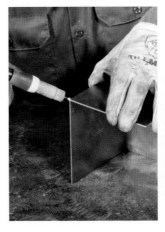

The next two pieces are easier to attach. You can hold the pieces, or use a magnetic square to keep them in pace. *Monte Swann*

FABRICATION AND ADDITIONAL EXERCISES

199

Tack near the corners. Try not to overlap the pieces. Open-corner joints are easier to weld than closed-corner joints. Half-open corners, which partially overlap, are not bad. If the pieces are not square, one end may overlap while there is a gap in the other end. *Monte Swann*

On larger cubes, it may help to make a small tack in the middle of the joint. Any gaps should be closed up at this time. Use clamps to pull pieces together. *Monte Swann*

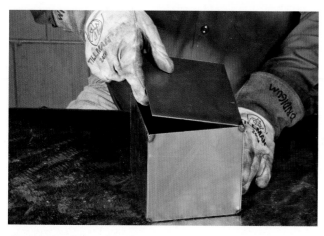

With only one tack on this piece, it is easily shifted into position. *Monte Swann*

Use soapstone to mark the order each joint is to be welded. *Monte Swann*

Add a little filler rod to the corners to build up weld metal. *Monte Swann*

Skip around on the cube. Don't make welds all in one area. This prevents excessive heat from building up in one part of your project. *Monte Swann*

Use filler rod to fill any openings. Filler rod can also be used on the whole piece. *Monte Swann*

Extra gloves come in handy to pick up large, awkward pieces of hot metal. *Monte Swann*

The finished cube can be made into a piggy bank. Two cubes can make a giant pair of dice. *Monte Swann*

WELDING ALUMINUM WITH GTAW

Welding aluminum is not any more difficult than welding steel, it's just different. Read through the GTAW chapter for information on the welding process, recommended electrodes, amperage settings, and how to prepare the tungsten for welding with AC. The intense temperature of the tungsten arc eliminates many of the problems encountered with gas-welding aluminum. There are a few considerations when welding aluminum with GTAW:

1. Aluminum will not change color when it is heated. Unlike steel, there are no telltale clues on how hot the base metal is until it becomes too hot.
2. A layer of oxides are always present on the surface of the metal. These oxides melt at 3,700 degrees—2,500 degrees higher than the melting point of aluminum. This oxide layer on aluminum (and some brasses) is like the shell of a hard-boiled egg. It takes more force to bust through, but once you do, little effort is required to get to the yoke. It takes more heat to melt through the oxide layer on aluminum. By the time you do, the base metal has been overheated. Often, the aluminum will melt or turn mushy underneath the oxide layer before a molten pool forms on the surface of the metal. Oxide layers can vary in thickness, depending on the type of alloy, structural shape, and storage conditions. For example, aluminum stored outdoors will have a heavier oxide layer.
3. Aluminum has a high level of thermal conductivity. Heat leaves the weld area at a rapid rate.

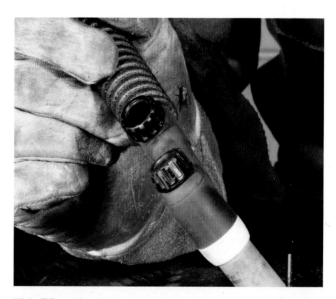

While TIG welding, this fly on my glove was attracted to the arc light. It is not uncommon for bees, wasps, and other insects to go directly for the welding arc. Kids, pets, and other people are also attracted to the light. Do what is in their best interest and keep them from getting arc flash. *Monte Swann*

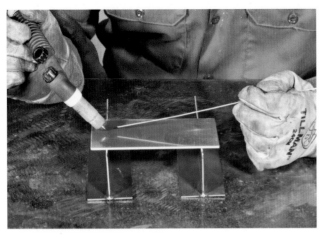

The torch and filler rod angles are the same for TIG welding aluminum or steel. *Monte Swann*

Form a puddle on the base metal first. Look for part of the metal to become a shiny liquid. The base metal will not change color; it will only look wet when melted. *Monte Swann*

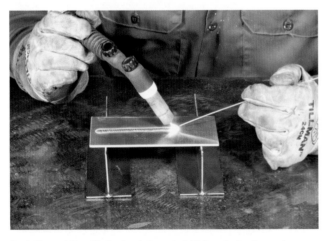

Push the puddle with the torch in a straight line. A circular or zigzag motion is not necessary. Keep the filler metal close to the molten pool, within a ¼ inch or less. *Monte Swann*

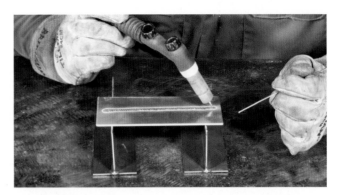

After the bead is finished and the arc is off, keep the torch over the solidifying metal for a few seconds. That way the shielding gas continues to protect the crater during this crucial time. *Monte Swann*

A lap joint in the horizontal position is shown here. Begin the molten pool on the base metal, first. Then, tilt the torch to melt the edge. *Monte Swann*

Add filler metal to the C-shaped weld pool. Add filler metal frequently and consistently for proper fusion at the root of the joint. *Monte Swann*

As you approach the end of the joint, decrease the amount of amperage. Heat will concentrate near the end of a weld. With aluminum, there will be no indicator (like a color change) before the metal melts through. Add enough filler metal at the end to fill in the crater. The amperage can be increased slightly to remelt the crater and fuse in additional filler metal. *Monte Swann*

A uniform travel speed and consistently adding filler metal produces beads of a uniform width and height. Use enough amperage for good fusion into the base metal. *Monte Swann*

4. Although aluminum has a lower melting point than steel, aluminum has a higher specific heat. This means it takes more energy to raise the temperature of aluminum one degree than steel. Think of it as a race between a bus (aluminum) and a muscle car (steel) with the finish line being the melting point of each metal. The bus is heavier and slower, but the distance to the finish line is shorter. The muscle car is much faster, but has to go a longer distance; just like it takes less energy to heat steel, but steel has a higher melting point than aluminum.

To weld aluminum, first clean off any oil, coolant, grease, dirt, or dust from the surface of the metal. The oxide layer can be removed by lightly brushing the joint with a stainless-steel brush, sandpaper, or plastic scouring pad. Use a brush dedicated to aluminum only, and brush in one direction. Applying too much pressure can imbed even more oxides into the metal; never use a power brush to clean aluminum. The oxide layer is also broken up by the reverse polarity portion of the AC wave. Making an adjustment to the AC balance can go a long way in helping to get through the oxide layer and not overheat the metal.

On sections greater than ¼ inch, the base metal can be preheated up to 200 degrees F. Preheat provides a sufficient amount of heat for adequate fusion to take place. It is possible to weld ⅜- or ½-inch-thick aluminum without preheat. With a large welding machine, enough amperage is generated to make a first pass on thicker metals. The first pass then acts as preheat for the fill and cover passes. Be careful not to overheat the base metal. Prolonged exposure to heat can reduce the mechanical properties of heat treatment in aluminum. Keep preheat temperatures below 400 degrees. Use a temperature indicator, such as a temp stick, to keep track of the heat input.

Build up the craters at the end of the weld beads. Concave craters are prone to cracking, especially with aluminum.

Aluminum takes more amperage to weld than steel because of higher thermal conductivity and higher specific heat. The use of AC, which only penetrates into the base metal during the straight polarity part of the AC cycle, also contributes to the need for a higher amperage setting.

CONCLUSION

Even with all the information contained in this book, we have only seen the tip of the iceberg. If anyone ever says to you, "I know everything there is to know about welding," they are not being honest. No one knows everything. No matter if you are welding for a living or just as a hobby, there are always new things to learn and new techniques in metal fabrication. The possibilities are endless; explore and educate yourself. Watch other people weld to see how they do it. Pick up information from other books, off the Internet, at your local welding suppliers, from manufacturers of welding equipment, and organizations such as the American Welding Society. Take a class. Get your hands on machines you may not have at home and learn different welding processes. Each welding process will teach you something new, and at the same time reinforce the overall principles basic to all welding: distance, angle, speed, and heat. Work safely and have fun. Welding is a skill learned with practice and guidance. Over time, welding will allow you to build projects that are functional, ornamental, or sculptural.

RESOURCES

BIBLIOGRAPHY

Finch, Richard. *Welders Handbook,* HP Books, 1997

Sacks and Bonnart. *Welding Principals and Practices,* McGraw Hill, 2005

Moniz and Miller. *Welding Skills,* American Technical Publishers, 2004

Jefferson, T.B. *Metals and How to Weld Them,* The James F. Lincoln Arc Welding Foundation, 1990

The Procedure Handbook of Arc Welding, The James F. Lincoln Arc Welding Foundation, 2000

Fabricators and Erectors Guide to Welded Steel Construction, The James F. Lincoln Arc Welding Foundation

Welding Inspection Technology, American Welding Society, 2000

Welding Handbook Volume 2, American Welding Society, 1991

Guide to Nondestructive Examination of Welds, American Welding Society, B1.10:1999

Standard Welding Terms and Definitions, American Welding Society, A3.0:2001

Structural Steel Welding Code, American Welding Society, D1.1/D1.1M: 2006

LINKS

www.aws.org

ANSI Z49.1 Safety in Welding, Cutting and Allied Processes

www.millerwelds.com

Click on "Resources" and "Improving Your Skills" to find page with helpful articles about all types of welding along with a welding dictionary, welding books, and more.

www.lincolneletric.com

Welding equipment manufacturer

www.norco-inc.com

Lincoln Electric informational articles.

www.esabna.com

Click on "Education" and on "ESAB University" to get to listing of helpful handbooks, courses, and material property listings.

www.harrisproductsgroup.com

Information on brazing and soldering.

http://manufacturing.stanford.edu

Click on "How everyday things are made" to view short narrated films showing manufacturing processes.

ACKNOWLEDGMENTS

Thank you very much to the following people, organizations, institutions, and businesses. Without all of them this book would not have been possible.

My wife ,Gina Bridigum, for encouragement, inspiration, and assistance in writing and editing; Helen Henning for assistance with editing; Dave Fitzgerald and Rick Dahlstrom for teaching me how to weld; Quay Grigg for teaching me how to write; and to Dwight Affeldt for his technical assistance with this book and making critical contributions to the content.

Others who provided valuable technical assistance include: Kim Munson, Machine Tool Technology program at MCTC; Greg Skudlarek, HVAC & R program at MCTC; Pam Lesemann, Oxygen Service; Mace Harris, Twin City Oxygen; Ryan Salmon for his polished and etched weld specimens and "welding time" project; Monte Swann Photography for an excellent job working with me to capture the essence of welding; Rob Lindgren for providing two key technical drawings; Minneapolis Community and Technical College for allowing the use of the welding lab for photography; the American Welding Society for allowing us to use a great number of their figures and tables from key publications (AWS D1.1/D1.1M 2006 *Structural Steel Code, Welding Handbook Volume 2, Welding Inspection Technology,* and *Standard Welding Terms and Definitions).*

ABOUT THE AUTHOR

TODD BRIDIGUM CWI/CWE

Certified welding instructor Todd Bridigum was drawn to the practical and artistic side of welding and metal fabrication. After graduating with a bachelor's degree from Hamline University, he decided to enroll in trade school and worked at many fabrication shops doing production welding and learning trade secrets from the pros in the field. When there was a downturn in industry after 9/11, he began to teach welding part-time and later became a welding and metal fabrication instructor at Minneapolis Community and Technical College. Along with teaching traditional students, he has run customized classes for the Minnesota Air National Guard, 3M, Toro, Anchor Block, Laborers Union, Construction Electricians Union and the light rail division of MNDOT. In a study funded by the CDC on the causes of workplace amputations, he performed assessments and training in many metal fabrication shops as part of the research into this area by Dr. David Parker. In his free time, Bridigum designs and builds printing presses used for fine art intaglio printing from etched and engraved copper plates.

INDEX